A JOB WITH
ROOM & BOARD

Memories of an Early Montana Forester

A JOB WITH
ROOM & BOARD

Memories of an Early Montana Forester

JOHN B. TAYLOR

Foreword by JOHN N. MACLEAN
Edited by JOHN C. FROHLICHER

2015
Mountain Press Publishing Company
Missoula, Montana

Library of Congress Cataloging-in-Publication Data

Taylor, John B., 1889-1975.
  A job with room and board : memories of an early Montana forester / John B. Taylor ; foreword by John N. Maclean ; edited by John C. Frohlicher. — First edition.
      pages cm
  Includes index.
  ISBN 978-0-87842-639-3 (pbk. : alk. paper)
  1. Taylor, John B., 1889-1975. 2. Foresters—Montana—Biography. 3. United States. Forest Service—Officials and employees—Biography. 4. Montana—History—20th century—Anecdotes. 5. Deerlodge National Forest (Mont.)—History—Anecdotes. 6. Outdoor life—Montana—Anecdotes. 7. Montana—Social life and customs—20th century—Anecdotes. 8. Montana—History, Local—Anecdotes. I. Frohlicher, John C., editor. II. Title. III. Title: Memories of an early Montana forester.
  SD129.T36A3 2015
  634.9092—dc23
  [B]
                                    2015004958

PRINTED IN THE U.S.A.

MP Mountain Press
PUBLISHING COMPANY
P.O. Box 2399 • Missoula, MT 59806 • 406-728-1900
800-234-5308 • info@mtnpress.com
www.mountain-press.com

To Catherine, Elsie, Dora, and Ellen,
who have heard all this before.
Now written for my grandchildren and others
interested in this early history.

Forest ranger on the Gallatin National Forest in 1938. —Photo by K. D. Swan, US Forest Service

CONTENTS

John B. Taylor (June 9, 1889–January 17, 1975) looking over Moose Lake in Granite County toward the Anaconda-Pintler Wilderness. —Photo by Catherine H. Taylor

FOREWORD

The world that John B. Taylor describes in northwest Montana at the turn of the past century was a mix of Eden and Eden after the Fall. In Missoula, upstream copper-mining operations had so fouled the Clark Fork River that it could not support fish or other aquatic life: the river smelled dead as it passed through the city, where it received another dose of filth. Outside town, rapacious timber and mining operations denuded the hills; wolves, grizzly bears, mountain lions, and coyotes were hunted relentlessly; game laws were few and mostly ignored.

When Taylor set out on a survey trip for the Forest Service in 1907, however, mapping the Swan River Valley, he stepped into a virgin wilderness of breathtaking freshness and beauty. His account of his experience on crews of young Forest Service recruits—a mix of graduates from "back East" schools and local boys—could be just another nostalgic wander across a landscape long disappeared. Fortunately for the book, the land, and the reader, it is more than that. Taylor's account is a valuable historical record of the first European parties to go into the Swan and Clearwater drainages in a systematic way, documenting the timber resources.

More relevant to the modern reader, Taylor describes a region that against all odds has retained huge reserves of exactly the kind of landscape, flora and fauna, and possibility for adventure that Taylor experienced, most significantly in the Bob Marshall Wilderness, the Mission Mountains Wilderness, and the Flathead and Lolo National Forests. Taylor gives us hints of how this came about at the grassroots level. The East Coast and Missoula boys on the survey crews could have fallen into the trap of mutual resentment, separated along regional and educational lines, and wound up in all-too-familiar gridlock. Instead, according to Taylor, the Westerners learned a greater environmental awareness from the Easterners, who in turn took lessons in woodsmanship. It's the stuff of a quintessential American education, half formal book learning, half hands-on in the outdoors.

Taylor's narrative jumps around a bit after the opening chapters, recording his long and successful career in the Forest Service, where he rose high in the supervisory ranks. The book has an underlying theme, however: the gradual awakening of America's environmental conscience, as seen through Taylor's own evolving awareness of and love for the natural world. You need both sides of the equation, the hard science and the loving heart. After many stops along the way Taylor winds up in Glacier National Park, where he offers a guided tour that should be made available at the park's visitor centers.

Taylor's account ends in the late 1960s, before the great successes of the environmental awakening of which he was an early part, such as the massive and successful cleanup of the Clark Fork River, and the broad challenges wrought by a warming planet and growing population. The principal lesson of the book, however—to respect learning and the land—is ageless.

Because I have spent time since infancy in the Swan-Clearwater region, Taylor's vivid accounts of his time there are the most compelling chapters, calling up not just memories of what was, but also of what still is. I have a son who has a special love for the Mission Range, and he and I have hiked and camped there since he was old enough to carry a pack. One recent early fall, the best time in the mountains, we walked up to one of those emerald-bottom crystalline lakes that are sprinkled along the timberline at the foot of the Missions' glacier-capped peaks. Taylor's survey crews had to bushwhack straight up from the valley floor to the timberline, over and over, and it was a rough go. My son and I followed a foot trail to the lake but then had to bushwhack another hour to get around to its head. We saw no other humans. We passed spruce trees of astonishing girth, there since before Lewis and Clark. The incredible sweetness of the air, its chill clarity, makes you catch your breath. There's less of it around now, the experience of raw nature, but it's still there.

—JOHN N. MACLEAN

PREFACE

This is not a history of the US Forest Service but of men who have served with the loyalty and dedication of crusading conservationists. With a sudden surge of resolution I have recorded these incidents to appease the friends who continually urged me to do so.

This is not fiction. The people are real. The names are real. Incidentally, some of these people went on to staff university forestry schools throughout the United States or state and private conservation departments, or they advanced to high positions in the federal service. These stories have not been published before. They provide the chinking or caulking in the cabin of history we are building.

—John B. Taylor, 1969

ACKNOWLEDGMENTS

- John C. Frohlicher, whose dynamic force achieved the tapes on which these memories were originally recorded, and who edited the first copy.

- Harold G. Merriam and A. B. Guthrie, both for encouragement and counsel.

- The US Forest Service for its excellent maps.

- Many friends who pushed for completion of this project.

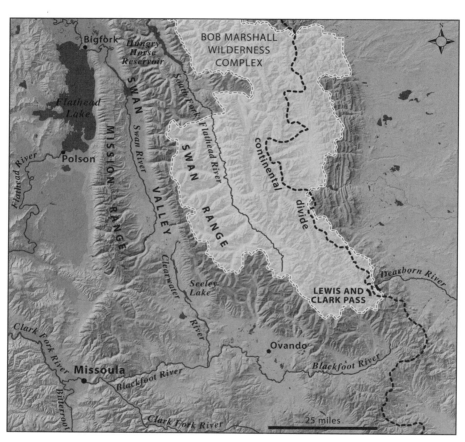

The region of the Swan Valley and other points north and east from Missoula, Montana. Note that the wilderness areas of the Bob Marshall Wilderness Complex were not designated until the mid to late 1960s.

Blazing a Trail

Back in 1907, more than fifty years before the nation as a whole became concerned about environmental problems, a series of events decreed that I would become a conservationist. Setting the stage for this, prodded by the American Forestry Association, Theodore Roosevelt, and many other forest conservationists, Congress passed an act on February 1, 1905, transferring the forest preserves from the Department of the Interior to the Department of Agriculture. The US Forest Service was thereby created out of the Bureau of Forestry. This started a sort of tidal wave that caught me and thousands of other young men, redirecting our lives to exploring and mapping wildernesses, fighting forest fires, and otherwise carrying on the crusade for forest conservation. Even as now, young men were called upon to play a dynamic part in our nation's history.

As an eighteen-year-old boy, preparing to enter the University of Montana with the goal of becoming a teacher of Latin and Greek, I was one of those affected. My home was in Missoula, Montana, the northwestern part of the state, where destructive logging followed by unrestrained forest fires adjoined glorious wildernesses. The Clark Fork of the Columbia River flowed through our community, but it was without fish or other life, for the mines and smelters of the Butte mines, about 120 miles to the east, had polluted it with copper salts and with tailings from the mills. Around Butte and Anaconda in Montana, and in the nearby Coeur d'Alene mining district in Idaho, smelter fumes had almost completely denuded the surrounding forests and mountains. Stockmen with both sheep and cattle competed for the open range and overgrazed it ruthlessly. The few game laws were disregarded openly. I had grown up in this environment and except for other circumstances would have continued to accept such destruction as normal progress. We assumed it to be axiomatic that ruthless war for extermination of such predators as bears, coyotes, wolves, and mountain lions was a proper and virtuous policy.

But plans can be changed abruptly. And so it happened with mine when early in the spring of 1907 a severe financial panic bit the country. It became impossible even to cash a check. To lighten the burden on my parents, I quit school early and looked for a job. After a few weeks as a general handyman and flunky on a railroad survey crew, I was hired by the US Forest Service for the summer. Actually, my father located the job, for I was working about fourteen hours a day and had no chance to seek other employment. He informed me that "the forest reserves were hiring boys," that the work would be out in the open, and that, as it was a government job, I probably would not have to work very hard. So, at the first opportunity, I got off work early and went down to the old First National Bank building in Missoula to put in my application.

At this headquarters of what became Region One of the Forest Service, responsible for more than twenty million acres of national forests, I found the staff to consist of the chief inspector, E. A. Sherman; one assistant, Elers Koch; and one woman clerk whose name I did not get. Mr. Sherman appeared to me to be an old man, though I now realize he was about thirty-five years old. Koch was a recent graduate of Yale School of Forestry and in his midtwenties. At that time there were few old men in the Forest Service.

Mr. Sherman subjected me to a brief interview and hired me at the rate of $50 a month "with room and board." There was lots of room, some eight million acres of the Lewis and Clark National Forest, embracing what is now divided into several forests together with what later became Glacier National Park. Elers Koch gave me instructions for reporting to the job. With five other students, I was to join a party that was mapping the Swan River Valley, then a virgin wilderness lying west of the present Bob Marshall Wilderness. They were also making a cruise, that is an estimate, of the timber there. We were directed to take with us our field beds and everything except food that we would need for the summer. There would be no chance to buy anything. We were also told that we would have to pack our personal effects on our backs, and to "make it light." Of course, that was not a new idea to boys living in Montana. Every lumberjack, cowboy, or other worker went equipped with what was variously called his "turkey" or "bindle," a bedroll in which were wrapped all his personal possessions.

The 130-mile trip to report to the job was typical of travel in Montana in 1907. We left Missoula on the evening of July 4 for a thirty-mile train trip to Ravalli, a railway station on the south side of the Flathead Indian Reservation, where we spent the night in the crude, wooden hotel. Along the railway siding in Ravalli, we observed with special interest the big corrals built of heavy logs with loading chutes currently being used by two well-known Indians, Charlie Allard and Michael Pablo, in shipping their buffalo herd to Alberta, Canada. We knew they had built up one of the few remaining herds in the country and had offered it for sale to the United States. Ironically, having killed off almost all of the huge herds of buffalo, our country was unwilling to pay for the preservation of the few remaining, and Congress had refused to appropriate funds to buy them. The Canadian government showed greater vision and promptly purchased a large number. Belatedly, the United States bought what remained to form the nucleus of the herd, which now occupies the National Bison Range on the Flathead Indian Reservation northwest of Ravalli.

Next morning, on July 5, we left Ravalli on a Concord stagecoach drawn by six horses for the forty-five-mile trip north across the reservation to Polson at the foot of Flathead Lake. The road was an unimproved wagon track winding among the rocks. To the east rose the rugged, snow-capped peaks of the Mission Range. Our summer work would be on the other side of this mountain barrier. Having climbed over a low ridge of open hills and down to the early Jesuit Mission at St. Ignatius, we changed teams for the next section of ten or twelve miles to the old trading post at Post Creek. On and on we rolled, with additional changes of horses, winding among a series of glacial ponds to Ronan, the merchant center of the reservation, over the high glacial moraines overlooking Flathead Lake and the little town of Polson. The trip culminated in a triumphant

A typical Concord stagecoach circa 1907, using four to six horses.

3

dash down through the town to the dock, where the stern-wheel steamer *Klondike* awaited our arrival. The forty-five-mile trip had taken six hours, using four different teams of horses, at an average speed of nearly eight miles an hour—good time for stage travel.

The Indian reservation through which we traveled had not been opened to white settlement at that time and almost the only whites on the reservation were those at the mission and a few permitted traders and merchants at Ronan and Polson. Most of the Indians still wore moccasins and blankets, the men in faded, worn Levis, and the women in calico skirts. We saw one old Indian at Ronan who had evidently found his overalls uncomfortable, for he had cut out the seat to convert them into something resembling chaps. Then he had added a breechclout. Picturesque tepees dotted the landscape while the few occupied dwellings outside the towns were crude frame shacks, usually with tepees nearby.

The *Klondike* sailed about 1:00 p.m. from Polson to our destination, Big Fork, where the Swan River flows into the lake at its northeast corner—about thirty miles. However, the *Klondike* was the chief carrier of freight, passengers, and mail along the west side of the lake. It visited Indian settlements at Big Arm and Elmo, stopped briefly at Dayton, and even crossed to an isolated logging camp on the east shore to discharge supplies. There, after the supplies were unloaded, four men brought aboard the body of a lumberjack loosely wrapped in canvas and deposited it on the foredeck. Society had little more concern about conserving its manpower than it had for its forests. Employees' liability and safety measures were concepts not adopted until later. But some disposition had to be made of a body, and this one was being shipped to Kalispell for burial, probably at county expense. It was after 6:00 p.m. when we disembarked at the little settlement of Big Fork. There we were met by a young forester, newly arrived from Washington, DC, who took charge of the party. We ate dinner at the primitive hotel, but since we had our bedrolls with us, we didn't waste money on hotel rooms. Instead, we slept on the ground at the edge of town. Next morning, after breakfast (the last expenditure of the summer), we shouldered our packs and started on the last lap of our journey to find camp, somewhere about thirty miles up the river.

The trip to camp took nearly two days, for we were soft after a winter in school, and unaccustomed to hiking over rough trails with packs on our backs. For six or eight miles we followed a crude road; then we obtained a small boat for the trip up Swan Lake. From the head of the lake there was only an old Indian trail through a magnificent forest of large yellow pine (ponderosa), larch, and Douglas-fir. Indians did not clear their trails, and this one wound through the timber, twisting and turning to avoid fallen logs and thickets of brush. It was hot, and clouds of mosquitoes feasted on our unprotected faces and hands. There were flowers in profusion. At openings, whitetail deer, unaccustomed to men, watched us with curiosity before bounding into the shelter of the timber. As the sun sank behind the Mission Range, the evening turned cool and the mosquitoes ceased their persecution. Our camp was a bivouac in a parklike meadow beside a clear stream. We cooked the food we carried with us and spread our blankets in the open under a clear sky. This was the life we were to lead for two and a half months. The next afternoon, after missing the trail two or three times, we reached our main camp. In three days we had completed a journey that now takes slightly over two hours.

At this stage, we six students had little idea of what we were to do or why we were there. However, next morning we learned that our party was to make a rough map and timber estimate of some half million acres, covering the entire drainage of the Swan River and of the Clearwater River, a tributary of the Big Blackfoot River to the south. We worked in parties of four men each, running lines east and west by compass and pacing* and recording by species and diameter the trees on a strip one chain (four rods or 66 feet) wide. Each line started from a mile post established by a survey crew on the main trail and was continued until it reached timberline on either the Mission Range to the west or the Swan Range to the east. The men carried packs containing two blankets for the four men, together with food sufficient, we fondly hoped, to last out the trip. The line went straight through every thicket of willow or devil's club (a thorny brush), through tangles of fallen timber from old fires, across streams and swamps, and up rough mountain canyons. As night approached, we camped at some stream, cooked a supper of salt pork, rice, and a hard cracker called pilot bread. Frequently, this diet was supplemented with fool hens (spruce grouse), which rely on protective coloring and perch quietly on a branch. We killed them with sticks and stones. Our bed was a mattress of fir boughs. If the night was cold or rainy, we erected a little square of canvas for a makeshift shelter. Then we spread our two blankets over the four of us, keeping our clothes on. If the weather was cold, as it often was up in the mountains, we built a fire at the foot of the bed. This was our trail camp.

---

*Following a straight line by watching a compass and estimating the distance covered by counting the paces taken.

The main camp was relatively luxurious. There we had tents and a cook for the crew of about twenty men. Supplies were brought in by packhorses from Big Fork or, later in the summer, from Ovando, in the Blackfoot Valley. After running a line for a week or ten days, we were permitted a day of rest in the main camp. The personnel of the crew was typical of the men who built the Forest Service, a mixture of young foresters, graduates of the forestry schools at Cornell, Yale, or Biltmore, and of western men and boys. The college foresters usually had spent a year or so in the Washington office, where they were inspired by none other than Gifford Pinchot, first chief of the US Forest Service. They were a group of crusaders for forest conservation, though completely unskilled and inexperienced in woodsmanship. We westerners provided the practical woods skills. Incidentally, we absorbed from those foresters a new concept of forest protection and conservative use.

The chief of the party was Karl Woodward, a graduate of Cornell University forestry school, where he had studied under that great

Swan River camp in 1907. Karl Woodward, chief of the party, is standing with tent in background; Cooney, a Northern Pacific Railway man, with arms folded in his Teddy Roosevelt Rough Rider outfit; Dick Groom, seated, in profile.

German forester and educator, B. F. Fernow. Many years later I was to meet him again when I was on a lecture tour of eastern forestry schools. He was then dean of the forestry school at the University of New Hampshire. He and his wife, a western girl, entertained me at dinner in their charming old colonial home, and we relived that first year. Junius Benedict, who later became supervisor of a large forest, was, I believe, from Biltmore School of Forestry at Asheville, North Carolina, which was conducted on the Vanderbilt estate by another German forester, Carl A. Schenck. Bill Piper was a Yale forester. He later returned to the Midwest, where he was supervisor of the Huron National Forest in Michigan. About 1948, as personnel officer of that region, I processed his application for retirement.

My first day's work was with a young forester, Robert Y. Stuart. There was a rumor in camp that he was not in good standing and might be fired—why I do not know. At any rate, the rumor must have been exaggerated, for the next time I met him, many years later, he was the distinguished chief of the Forest Service.

The westerners of the party were a diversified lot. First was Mr. Cooney. Alone of the crew, he continued to be addressed as "mister." He was a short, stocky man of middle age, extremely cross-eyed. It was disturbing that one never was sure whether he was looking at one or at someone else. Perhaps he looked in both directions at the same time, for there was little he missed. Mr. Cooney was neither a Forest Service employee nor a forester but a land agent representing the Northern Pacific Railway Company, which owned most of the odd-numbered sections of land by virtue of the land grant it had received to subsidize the railroad construction. The company paid part of the expense of the survey to obtain information on its grant lands, and there were hopes that through exchange both the company and the Forest Service could consolidate their respective ownerships. Mr. Cooney had been a member of the Rough Riders in the Spanish-American War under Theodore Roosevelt, whom he worshipped. Perhaps because of that memory, he wore the khaki trousers, canvas puttees, and campaign hat of that organization on this job.

Dick Groom was the oldest man in the party, possibly sixty years of age. A slender, wiry man, he was indefatigable. Dick was about as much a part of the indigenous fauna as were the deer, or even the mosquitoes, which never seemed to settle on him, preferring the tenderer meat offered by the younger men. No doubt he was hired for his knowledge of the country and for his woodsmanship. His home was a neat log cabin near the head of Swan Lake. If he had any other home or relatives, he never mentioned them. It was said that Groom was not his true name, that he fled eastern Montana when detected carrying a running iron he was using to brand cattle claimed by others. One did not pry into the backgrounds of hermits like Dick. He could build a fire quickly under the worst conditions, knew just where the little shelter should be placed to avoid the wind, how to catch fish or fool hens. Dick never gave us less-experienced men direct instructions but taught us by indirection and example. I suspect he derived a lot of amusement from the eastern college men. He would watch them with amazement as they tried to coax heat from a smoky fire of half-rotted spruce; then he would saunter out, collect some stumps of fire-killed lodgepole pine from which the less resinous trunks had rotted, together with chunks of thick fir bark, and shortly would have a hot, smoke-free fire. At least, it was smoke-free for him; his bed was always so placed that any smoke drifted the other way, preferably where we others had made our beds.

Finally, there were the western boys, including the six of us from Missoula. Raised on ranches or in small western towns, we were experienced mountaineers compared to the eastern foresters, and compared to Dick Groom, rank tenderfeet. Of the six of us, I was the only one ultimately to follow the profession of forestry. Bernie Kitt, Bill Vealey, and Arthur Bishop all became construction engineers. They and others like them are the ones who have made it possible today to make the trip in two hours, which required three days for us in 1907. Bernie, for example, later was a location engineer on the construction of Going-to-the-Sun Highway across Logan Pass in Glacier National Park.

This was the party, about half westerners and half eastern college-trained foresters. The foresters were, as I have said, ardent crusaders, partisan forest conservationists who were sure the country would run out of timber presently. They preached the doctrine of forest conservation constantly to the rest of us. Their forestry was strictly of the German school, taught them by Fernow, Filibert Roth, and Carl A. Schenck, all Germans and graduates of German universities. If the technology was ill adapted to our virgin forests and to the economics of our forest industries, the basic science of managing forests for continuous crops was sound. Moreover, most of these men had met Gifford Pinchot and had been inspired by his vision.

All summer we worked through that magnificent wilderness, virgin forest, unopened and unmapped. I remember camping one night up at timberline on the Swan Range in a saddle where we could see far to the east across the tangled canyons and ridges of the drainage of the South Fork of the Flathead to where the main continental divide thrust up its jagged peaks. At that elevation there was no big or thick timber and, with only a few scattered half-dead trees, no chance of starting a forest fire. We fired up a dead whitebark pine log to warm us as we lay on the rocks. It was cold at that altitude. We roasted on one side while we froze on the other. Some puzzled mountain goats settled down on a ledge above us where they could watch us.

When dawn arrived, all the jumbled canyons and lower peaks of the South Fork were veiled in fog. Far to the east, the higher peaks of the divide thrust their tops through the mist and turned red. The sun sent streamers through the gaps between the peaks, hitting the fog banks in vulnerable places to create a boiling ferment. As the sun rose higher, it chased fugitive masses of fog up the darker canyons, gradually dissipating

Mountain goat.

Measuring a big larch, 1907–08.

every remaining cloud until the entire drainage was bathed in morning light. Then the goats gave up their watch and departed to their alpine pastures.

Not all lines led to such scenery. Early in September, four of us started a line west from the head of the Clearwater River, a strip that would cross the divide back into the head of the Swan River drainage and reach far up into the Mission Range. We optimistically estimated that we could complete the strip and return to camp in seven days and took food accordingly. Benedict was the compassman, holding to a due west line by means of a box compass, pacing to measure distance, and making a rough map, which included readings of altitude at fixed intervals from an aneroid barometer. Bill Piper was tally man. He made notes on the type and condition of the timber and recorded the species and diameter of all trees over eight inches in diameter, as Bernie Kitt and I, working on each side of the line, measured them and called them to him. Each of us caliper* men covered a strip thirty-three feet out from the center line. Ultimately, draftsmen combined the rough map with others on parallel strips, and computations were made of the total stand of timber from Piper's notes.

---

*A caliper is an instrument used to measure the diameter of a tree.

Unfortunately, our guess as to the time this would require proved to be low. We fought our way, all burdened with packs of blankets, food, and cooking utensils, through thickets of small lodgepole pine, climbed over fallen, fire-killed timber, and forded Beaver Creek and the head of the Swan River. The line climbed up the slopes of the Mission Range, frequently crossing patches of slide-rock and over ledges. Though we were conserving our small supply of food and had killed a few grouse, we were nearly out of supplies. I realized that my father had been badly mistaken when he believed that one would not have to work very hard on a government job, but he was correct when he said we would be out in the open.

Evening of the tenth day found us camped in a dark stand of small, subalpine timber—spruce, subalpine fir, and whitebark pine, far up on the side of the Mission Range. This was in an area which is still wild and is being considered for recommendation to Congress for preservation as a wilderness.* We had completed our line, for we were beyond the upper limit of merchantable timber, but we were also virtually out of food. Supper consisted of two fool hens stewed with a small amount of rice, together with a cup of tea.

We decided not to try to retrace our line to base camp nearly twenty miles away, but to follow an easier—but longer—water grade route. At the first light of dawn, we shared our last rations, a small cup of rice for each of us and a cup of tea. Passing small pools already covered with ice, we made our descent to the Swan River, found a faint Indian trail, and wound our way through windfall, along the side of Lindbergh Lake (then called Elbow Lake) into the broad Swan Valley. There was no pause for lunch, since we had nothing to eat, so we made good time. Even so, the total distance must have been between thirty-five and forty miles by the time we reached main camp at twilight. I shall never forget dinner that evening. Game season had opened in September and someone had killed a deer. The cook warmed up a stew of venison and potatoes flavored with wild onions. I have never since enjoyed a meal more. That night we had the luxury of sleeping in tents where the frost could not form on our blankets; the next day we had nothing to do but to lie in the September sun, frequently arousing ourselves to eat a little more.

As we discussed the situation around the campfire we all agreed, easterners and westerners alike, that it was rather foolish to map this wilderness and to make a rough cruise of the timber. It could not conceivably be logged in our lifetime, we thought. Of course, we realized that the

---

*The Mission Mountains Wilderness was established in 1975.

11

Forest Service had to make some exploratory investigations of the forests over which it had been given jurisdiction. But certainly, we felt, this area was too remote to justify much study. How wrong we were! Now* there are standard roads along the main drainages with good gravelled roads leading off in every direction. Dude ranches and cafes, not to mention bars, line the roads. Tourists must be alert to dodge logging trucks. Lumbermen are rigorously opposing the withdrawal of any part of the Swan to be preserved as wilderness. One sees fewer deer. Instead, there are parties of tourists in strange costumes riding bored horses, or groups of Boy Scouts with packs on their backs duplicating our explorations of sixty years ago.

Mid-September found us camped by the shore of Seeley Lake, near the present ranger station site. Our summer job had ended. The chief of the party gave us sufficient food to supply us on our hike out, and we left camp for home, about fifty-five miles down the Clearwater and Blackfoot Rivers to Missoula. There was no portal-to-portal pay or transportation; we had to get back as best we could.

A few miles below Seeley Lake, we crossed the national forest boundary into an area of private land that had been logged by the old destructive logging methods. Fire had swept through the slash, leaving here and there some rugged old trees so decadent that the loggers had passed them by. It was a picture of devastation. Incidentally, that area has now fairly well recovered and supports a stand of vigorous young timber. Then we came to one of the abandoned logging camps—to filth of discarded clothing, tin cans, manure from the horse barns, and tumbled down buildings. We had returned to civilization.

That brutal exploitation of once beautiful country shocked and angered me. Of course, I had frequently seen the same thing before. However, this time I had been all summer in unspoiled wilderness, listening to the preaching of foresters. I saw the scene with a new understanding. Then and there I decided that spending my life helping to establish better management of our forests was a more worthwhile life for me than teaching Latin and Greek. I say this with no disparagement of the classics. I still find myself mourning at times that I haven't been able to give more attention to the humanities, but for me, live trees beat dead languages.

So we hiked home those fifty-five miles. We six university boys were eager to get back to Missoula and to enroll for the fall term at school. After all, we had worked for two and a half months at $50 a month, and

---

*In the late 1960s.

$125 was a lot of money in those days. We hadn't, of course, been able to spend a cent of it.

Though I continued at the University of Montana for four years, taking all the basic sciences and essential general education courses, I knew I would presently have to go east to a forestry school if I were to become a trained forester. At that time Montana had no forestry school. The problem of selecting a school was not great, for there were only a few universities offering full professional forestry training at that time. I shortly decided on either Yale or Michigan, and the choice between the two was finally decided on an irrelevant reason. As we sat around the campfire evenings, the college men naturally would reminisce about their college experiences. Yale, I learned, conducted a summer school in Louisiana. One evening, some Yale men started talking about the collection of snakes they had made at summer school. It must have been remarkable, including rattlesnakes, copperheads, water moccasins, and even a coral snake. I have a morbid fear of snakes. A little investigation revealed that Michigan didn't send its students into any snake-infested region. That settled it. I became a Michigan man and not a Yale product. I had started to chart my course.

Moreover, I was to be privileged to work and associate with men who became leaders in forest conservation on the national level, as well as in regional offices, and in universities throughout our country.

John B. Taylor ready to start on the job in May 1908.

ROLLING THE WILDERNESS INTO A MAP

My second season, in May 1908, I left school early to join a small reconnaissance and mapping party. The chief of the party was Jack Bentley,
fresh from Cornell University. It seems incredible that the Forest Service
would take a young college graduate with no experience and give him
such responsibilities. He simply got on the train and went directly to the
little flag station of Drummond, Montana, where he exchanged train travel
for a stagecoach to Ovando, where he met and took charge of the party.
This consisted of Fred Thieme* and me from the University of Montana,
old Dick Groom, Joe Waldbillig as packer, and Herman Rudolph as cook.

With the packstring carrying our outfit, we followed up the Blackfoot
Valley to Alice Fork, and up that to what is now known as Lewis and Clark
Pass. We followed the same route taken by Lewis on his return trip in
1806, even traveling in much the same way, on foot and with packhorses.
Squarely on top of the continental divide, we left the crude road and with
no trail, worked our way among the snowbanks along the divide to drop
down into the head of a small tributary of the Dearborn River. There, on
a pleasant May day, we established our first base camp, preparatory to
starting the mapping project.

The next morning we were astonished to find two feet of snow over
everything. For thirty days we never saw even a quick glimpse of the sun;
there was almost continual snow and rain. It was miserable. We couldn't
have gone out except by walking and leaving our outfit. Of course we
wouldn't do that and give up the job. We were in no danger, merely damp
and chilly. It was our packhorses that suffered for food, for grass was
covered by snow. At the end of the thirty days, the mountains were covered with two to four feet of new snow, added to the winter banks, and
completely saturated. Every small stream was already running bank full.

Finally, one morning in early June, we arose to a perfectly clear day
with not a cloud in the sky. The hot sun wreaked havoc with that saturated

---

*Fred Thieme years later became chief of engineering for Region One of the
US Forest Service.

A little canvas shelter in 1908. John B. Taylor at right.

mass of snow. The snow melted almost all at once, sending a mass of water into already swollen streams and tearing great gullies out of the mountainsides. Our little stream, which carried only a trickle of water in summer, became a torrent, dislodging boulders as large as a small house.

In Montana they still talk of the floods of 1908. For two weeks not a train ran on the Northern Pacific Railway between Butte and Missoula. Railroads were washed out, communities isolated, mail service ceased, homes destroyed. Not a bridge remained on the Blackfoot River, and the valley, which depended on freight wagons and stagecoaches, was marooned.

As for us, we could only guess as to what was taking place down in the valleys. But we were now free to map the small tract that we had expected to finish up in about two weeks when we originally went in there. When we completed the job, we headed back down the Black-foot, having to ford every little stream for they were still in flood. The North Fork, normally small, was a big, turbulent river. Our horses could cross, but there was no chance of our wading it. The packer took the pack string across and returned with bareback horses for the party to ride. It was my lot to ride double behind the packer on his saddle horse, an animal of uncertain disposition. As we climbed the bank on the far side, I was so careless as to slide back off the skirt of the saddle onto his rump. That did it. Never in my life, before or since, have I gone so high off a horse. There I was, well back on his hind quarters in a perfect position for him to throw me. For a while I wondered if I would ever hit the ground, but I did, and pretty hard too. It didn't do me any serious injury, however.

In Ovando, we renewed our supplies so far as possible, for they were in short supply. They depended on freighting for their supplies, and of course no freight wagons could get in. Then we went on fifty miles to Seeley Lake, up the Clearwater, across the Swan Divide, and to the Gordon Ranch in the Swan Valley. From there we headed east over Gordon Pass, attempting to go into the South Fork where we were to map another tract. But on top of the pass, we encountered about thirteen feet of old, hard-packed snow, and the descent on the northeast slope ahead was too hazardous for our horses. (The date was June 28.) We unloaded the packhorses and sent the packer back to the valley where he would remain until it became possible for him to cross the pass. Then we dragged our outfit on tarpaulins* down over the snow to a little lake where we pitched our camp among the snowdrifts.

Our admiration for Jack Bentley grew as he continued to surmount one difficulty after another, though until two months before, he had never been west of the Mississippi River or had even seen a packhorse. He was a good scout. In fact, he was a better scout, perhaps, than we were, because we used to laugh a bit at some of his tenderfoot tricks. He learned fast. Incidentally, almost twenty-two years after our summer together on this mapping party, in February 1930, I had dinner at his home in Ithaca, New York, where he was a full professor of forestry at Cornell University. He impressed on us the importance of the demanding routine of our job, telling us that only with those preliminary extensive maps would there be the required knowledge of the country from which to plan administration and protection.

Though we encountered discomfort from the elements at times—heat in summer, cold in early spring and late fall at high altitudes—we young men did not feel we were suffering undue hardship. As for danger, the hazards, such as there were, were familiar to us. It was part of our environment. We took suitable precautions to avoid injury and, aside from that, gave danger no thought.

---

*Canvas sheets used to cover packs or as tent flies.

Probably the worst thing about this life was extreme monotony at times. Several months out in the woods mapping and cruising* timber with a small crew of men represents day-after-day drudgery. Unless one had a real interest in forest conservation and in the phenomena of nature, he could wear out on the job and lose heart. Climbing through the brush and up and down the mountains in all kinds of weather became a serious threat to morale at times. I suppose it was more so, inescapably, to those boys who had come from eastern cities and who were not accustomed to this partially settled country and its isolation.

I remember one chap from Philadelphia who, incidentally, was raised in a strict Quaker home and whose nearest approach to profanity was to say "By Jove" or "Oh, dear." One can understand our surprise one day when he finally fell from grace. The packer had come in with a two weeks' accumulation of mail. We were sitting around in our tents reading our mail when Josh suddenly exclaimed, "Hell!" We were all shocked and asked him what the trouble was. He said, "Listen to this letter. 'How I envy you your wild, free life in the open.' Hell!" He had found that this wild, free life was merely day-after-day hard work.

Of course, we had experiences approaching adventure, defining adventure as hardship and danger. I remember mapping in the Swan Range, in what is now the Bob Marshall Wilderness. This was in 1908. We were mapping the high, rocky country where there was little timber, by occupying peaks, placing markers on them, then triangulating and intersecting and sketching in the topography.

Having occupied such a point and completing all the subsequent necessary work, we then wished to move over a matter of three-quarters of a mile or so to another prominent peak for additional work. There were two ways to go. We could drop down into an alpine cirque and then climb out through rocks to the peak, or we could take a short and relatively direct route over a hard, packed snowbank, virtually a small glacier. Not in any spirit of adventure, but just to save work, we decided to try to cross the glacier even though we didn't have ropes to rope together, or ice-axes, or any of the other equipment to take suitable precautions. Fred Thieme went ahead, gouging out steps as best he could, and the rest of us followed.

I was carrying a 36- by 40-inch plane table with a tripod, which handicapped me quite a bit, and I was bringing up the rear. I got about halfway across—out in the middle of the shelf—when the step broke under me,

---

*Cruising timber is when a forester estimates the number of board feet in standing trees.

*Pentagon Mountain and Dean Lake on July 7, 1928, in what is now the
Bob Marshall Wilderness, which was established in 1964.* —Photo by J. R. H.

and the tripod, plane table, and I all started down the glacier. It was, I
would judge, 800 to 1,000 feet down to where it levelled off and terminated in the talus slope at the foot of the mountain, which broke into a
small alpine lake in the cirque, in which floated small icebergs.

Down I went, wondering if I could manage to steer myself enough to
avoid hitting the large boulders stuck up through the snowfield at the foot.
Well, I did, and I stopped before I reached the lake, sliding out quite uninjured except for badly wearing out the seat of my pants. That was somewhat
of a disaster because there was no chance to replace them out there. I had
to mend those pants with part of a flour sack I got from the cook.

Another time on this same mapping project in the drainage of the Little
Salmon River, three of us, Jack Bentley, Fred Thieme, and I, climbed down
through cliffs into a box canyon, planning to climb up the other side to
occupy a peak. However, down in the canyon we were unable to see
far enough up the sides to select a feasible route. There were seams and
chimneys in the cliff, but when we tried them we ran into sheer cliffs.
Finally, we decided to return the way we had come, return to base camp
for a new supply of food, and start out fresh. But then we couldn't find the

particular crevasse through which we had come down. We'd go up one, thinking we were in the right place, only to come up against a sheer rock wall. Eventually, we had to go down the canyon to circle around.

This was an exceedingly rough, mountainous country. It had never been mapped except for sketching of the main drainages as they appeared to an engineer who established a triangulation point on a mountain about ten miles distant. Our box canyon led into one of these drainages which, according to the sketch, joined the stream on which our base camp was located. We decided to go down to the junction of the streams and back up to our camp, a circuitous route, but one which did not involve climbing over the ridges. However, night was coming on so we camped for the night. The next morning we started down the stream, but by noon we knew the sketch map was in error. I now know that this stream never did join that on which our base camp was located. They run into the main South Fork some fifteen miles apart. Probably some people will say we were lost. We didn't think of it that way; we knew where we were and where base camp was located, but we didn't know how to get from one place to the other by an easy route.

We now were very short of food, but of necessity we spent another night there. With fish lines, carried with us, we caught some trout and toasted them over a fire. The next day we climbed over the intervening ridge, down to the other drainage, and after a hard hike reached the base camp.

On that particular job in 1908, we mapped approximately one hundred square miles in an oblong strip along the east side of the Swan Range, an area that is still wilderness included in the Bob Marshall Wilderness. In all that vast area, we saw only one trace of human beings—the remains of an old Indian camp on the Big Salmon. This consisted of old tepee poles leaned against a tree, deer hair where they had made buckskin, and a sweathouse. Indians are in some ways more conservation minded than white people, and their camps are not as likely to show destruction and littering as those of the average white tourist. They would always lean their tepee poles that way if they didn't take them along so they would be available for future use.

The sweathouse consisted of a framework of willows bent over and thrust into the ground and just large enough to make a fairly roomy place for a man to lie down. This was covered with skins, the man crawled in nude, and hot rocks were shoveled in and sprayed with water to produce a steam bath, a sort of Turkish bath or Finnish sauna. These sweathouses were always on the bank of a stream so that at the conclusion of the sweat bath the Indian could literally jump out of the house into the cold water.

This old Indian camp constituted the sole sign that human beings had been in that country ahead of us, although I suppose there's little doubt that some wandering trappers and prospectors had been through there.

Rattlesnake Creek in 1909.

How Do You Name a Creek?

When the Forest Service took jurisdiction over the national forests, much of the area was utter wilderness with relatively few of the geographic features named. It was desirable, even imperative, that we have names so that we could refer to places specifically when bringing the lands under protection and management. It was part of the job to name these features.

One of the unfortunate developments in the early stages of this naming game was the tendency to name streams and mountain peaks after one's friends. For example, up in the Swan River and South Fork drainages, you will find on our maps today a Woodward Creek, a Woodward Lake, and I believe there's a Woodward Mountain. Karl Woodward was the chief of my first survey party—the first in this region—so we named some of the features for him. There's a Piper Creek named after Bill Piper, one of the college men on that party. He later became supervisor of a Michigan forest. Also, you will find Groom Creek and Bond Creek right at the head of Swan Lake, coming in from the east. Dick Groom and Ranger Ernest Bond, of course, were the ones honored. An abbreviation of Vanderwalker, the name of our packer, marks Van Lake and one or two other topographic features there today.

Cooney Creek on the Clearwater drainage preserves the name of a Mr. Cooney, who was a representative of the Northern Pacific Railway, which owned odd sections through that country. Because his company was also interested in learning more about its acquisitions, he was participating in this mapping and surveying project. He was quite a character but, nevertheless, a fine old gentleman. Also on the Clearwater drainage above Seeley Lake is Findell Creek, named for Elmer Findell, associated for many years with the lumber industry in the locality.

Jack Bentley, chief of our mapping party in 1908, drew the prize. You will remember that he met us at Ovando that historic spring, fresh from Cornell. In the one saloon in Ovando at that time was a picture titled "Sappho Emerging from the Bath." Over a month later, after we had crossed Gordon Pass into the Big Salmon, we headed up a side stream one day. Coming to a beaver swamp, we tried to figure how to get across without getting our feet wet. Finally, we found a narrow place spanned by a log. One of the other boys went first, then I followed, and then came Jack, who had not yet mastered the art of walking a log. He got out in the middle of it, teetered a while, and then went down "ker splash" into all that beaver muck and water. When he came out, I said, "Aha! Sappho emerging from the bath!" Since the stream was unnamed, we gave it the name "Sappho" on the map.

I hadn't thought of that incident for years, but in 1951, for the first time since 1908, I went down the Big Salmon again, this time with a party of trail riders of the American Forestry Association. Presently we crossed a little stream with a very neat enameled Forest Service sign: "Sappho Creek." That's how features got names.

Unfortunately, I have nothing named after me. Abuse of the practice of naming features after themselves led the Board of Geographic Names to issue a cruel order that prohibited the naming of any topographic feature for a living person. During World War I, my associates in the Division of Engineering in the Missoula office of the Forest Service reserved a fairly prominent peak north of Missoula to be named for me if I were killed. Fortunately, I missed perpetual fame. The mountain is now known as Stuart Peak. Now, if that had been in Canada—. And it is a beautiful mountain.

Up in Alaska, some thoughtful people crossed up the Board of Geographic Names. Axel G. Lindh, formerly chief of timber management in Region One before going on to higher responsibilities in Washington, DC, has his name on two islands in Alaska. A friend in that northern state named the most obstructionist mule in his pack string "Axel G. Lindh." The next year the mule died while contesting the right-of-way with a truck. In memory of the deceased mule, two islands were named: Little Axel Lind Island, which may have a hundred square yards of surface at low tide, and Big Axel Lind, with an area of about one hundred acres.

We named a peak up near Upper Holland Lake "Waldbillig Peak" after our packer, Joe Waldbillig, who had worked through that country a long time. That was not inappropriate, I think, because he was one of those whom one might designate as a pioneer of that country.

Some other places got names in peculiar ways. For example, there was Waterloo Ranger Station in French Gulch in the Big Hole, in the heart of an area that had been logged for mining timbers, convertor poles, logs, and cordwood for the reduction and refinement of Butte copper ore. Its name had an interesting history. When the area was turned over to the Forest Service, a French Canadian had a very disreputable roadhouse at that location. In order to have some semblance of legality to his occupation of the land, he filed a mining claim. The Forest Service promptly contested the claim in that he had never dug even a discovery pit and was doing no mining. The gold he was after was in the pockets of the wood-choppers and lumberjacks. The Forest Service won its contest, thereby evicting Napoleon Tessier. Needing some sort of a building to house the foresters there who administered timber sales, we appropriated the buildings. What was more logical than to call this Waterloo Ranger Station, since that was where Napoleon met his Waterloo?

Some of the names now on topographic features are, in a sense, euphemisms. The early mountain men often gave these places names that, at the very best, would have to be termed indelicate or even obscene. For a mild example, out west of Missoula there's a peak known as Squaw Peak. But from the early days of the settlers here, before the turn of the century, it was Squaw Tit Peak, which in English is indelicate. The name Teton National Park, which, of course, means exactly the same thing in French, doesn't seem to offend anyone. However, we dropped the second word in the old name, and it is now simply Squaw Peak.*There are many other names in this area that survive in purified forms from the original vulgarisms, which is perhaps just as well.

There's a peak on the Deerlodge National Forest that, due to its shape, was known to the ranchers as Horsecock Peak. Just west of the peak on Jerry Creek, a man raised horses and he had a very fine Kentucky saddler stallion. He was, incidentally, a rather interesting old gentleman. His name was Billy Delano. He was a distant relative of Franklin Delano Roosevelt, but he pronounced the name De-la-no with the accent on the second syllable. When it came to putting the name of the peak on the map, no one wanted to record it as Horsecock Peak, so someone suggested we name it after Billy Delano's stallion, and Starlight Peak it became. Starlight Peak it is today.

Over in the Clearwater National Forest, there's a marshy, wet area called Wietas Meadow. I read in a newspaper that it was named after

---

* The peak has since been renamed Ch-paa-qn Peak, which means "shining peak" in Salish.

25

Herbert H. Kuphal, John B. Taylor, and Henry Kuphal up the Rattlesnake in 1910.*

an old German trapper. I'm not prepared to say that isn't so; possibly it is. But the area wasn't known as Wietas Meadows in my youth. Its name consisted of two words, and anyone can readily imagine what those two words were. I doubt very much that there ever was a German trapper named Wietas; somebody merely purified the original vulgarism applied to that area.

Even as a boy, long before I knew how creeks and mountains were named, I was bemused by some of the characters whose names finally got on the maps. Up Rattlesnake Creek, for instance, some twelve or thirteen miles above Missoula, a little side stream called Beeskove Creek comes in. That was named for an interesting old character, Coyote Bill, whose real name was K. F. W. Beeskove. He was an old buffalo hunter who had his cabin near there. He had a team of mules, poached deer, and managed to eke out a living cutting cordwood for Missoula's fuel supply. He was at sword's point with his nearest neighbor, Arthur Franklin, who

---

* Herbert H. Kuphal became an engineer for the Bureau of Public Roads, now the Federal Highway Administration. Henry Kuphal received a silver medal from the Carnegie Hero Fund Commission in 1911.

lived approximately a mile beyond at the site of what is now the Franklin Guard Station.

Arthur Franklin was an entirely different sort of character from Coyote Bill. A highly educated man, he had a neat and exceptionally large cabin, beautifully constructed. In his exceptionally large library were books of the Greek philosophers in the original Greek, and a lot of Latin volumes. I saw him sitting in his home one evening reading the *Odes* of Horace. He was a cultured man and he was also a civil engineer, though he did not practice his profession. During the time I knew him, he had no apparent livelihood or source of income. Like a scattering of men through the west, he was a remittance man—one who received regular checks from his family in the east. He always seemed to have plenty of funds on hand.

Coyote Bill told us that Arthur Franklin was tied in with a gang of horse thieves, known as the J. C. Campbell horse thief gang, and that he housed them when they were collecting a few horses around Missoula. (There's no question that there was considerable horse stealing going on.) Whether or not Franklin was a party to this, Coyote Bill believed he was, and so did many other people.

Both men were dead shots; Franklin with a six-shooter, and Bill with his old .45-90 buffalo gun. I have seen Coyote—just to show us youngsters what he could do—shoot a squirrel out of a tall pine tree, which with a gun of that sort is quite a feat. I saw Franklin, while fishing, catch his fly in a branch of a tree across a little fork of the stream, take out his six-shooter, and shoot the limb off. Yet these two crack shots met and exchanged shots two or three times and neither ever hit the other. I could never understand that.

However, Coyote Bill was indiscreet in his reference to horse thieves. At least that was the presumable reason why some other men waylaid him. Franklin was not mixed up in this, but somebody beat the poor old man terribly. I saw him come to town past our home, which was in the edge of town on the Rattlesnake road. He was in his old wagon hauled by his mules, virtually blind from the swelling of his face. His jaw was broken and he had other injuries to keep him in the hospital two or three weeks.

When he finally got out of the hospital, he went back to the creek; but his home wasn't there. They had burned it down. So Coyote went on the warpath. (He had told me he was part Indian.) He started hunting, and when he found one of the men who had beaten him up, he shot him. He was duly tried and convicted. Undoubtedly, it was a case of premeditated murder. However, the judge apparently felt there was great provocation, for he gave Coyote only six years in the penitentiary from which he was paroled after about three years. While he was in prison, the prison chaplain converted him, and he came out deeply religious. He went over to the Flathead Reservation on the Jocko River, built a cabin, and lived out his remaining years praying regularly several times a day. His name is perpetuated on our maps, but few know the origin of the name or why Franklin Guard Station is so named.*

The Forest Service also acquired another ranger station from that horse thief gang near Bearmouth, about forty-odd miles east of Missoula. A man known as Red Pepper Jack had a cabin there on a squatter claim, that is, on land on which there was no filing and to which he had no title. He was alleged to serve the gang as a stopping place when they were taking horses through that part of the country. When the gang was finally broken up, Jack was convicted and sent to prison. The Forest Service then appropriated his abandoned home as a ranger station. That may seem high handed, but Congress was not making large appropriations for us in those days, and any loose properties of that sort on the forests were simply appropriated. If not luxurious, they were better than tents. However, Jack's name will not be perpetuated. Red Pepper Jack Ranger Station has long been abandoned and forgotten except by a few old timers.

That particular horse thief gang was reputed to work down as far as Pocatello, Idaho. It was a common joke that they created a demand for horses in Missoula territory by stealing horses, which had to be replaced. The replacements were horses stolen around Pocatello, thereby creating a demand in Idaho for the horses stolen in Missoula—a nice exchange, back and forth between the states.

There was another member of that same gang, a fellow by the name of Brady, whom Sheriff Wyman of Granite County with a deputy, Frank Morgan, went out to arrest. Wyman finally learned that Brady was holed up in an old cabin on a branch of Rock Creek, a place that could be reached only by saddle horse. They went there, and in the early morning they attempted to arrest Brady, who promptly reached for his gun and refused to throw up his hands. They brought him out, draped over a packhorse.

There was quite a lot of excited, public furor about the breaking up of this horse thief gang, which had contributed its share of the place names in this part of the country.

---

* The upper Rattlesnake is now part of the Rattlesnake Recreation Area and Wilderness, established in 1980. The guard station no longer exists, but a bridge across the creek is called Franklin Bridge.

WILD ANIMALS WE MEET

On these summer reconnaissance trips in unexplored country during the early 1900s, the native animals, bears especially, were always of great interest to the boys from the east. Some of them had been filled up with stories depicting bears as very dangerous. Actually, black bears are rarely dangerous; silvertips, yes. A silvertip grizzly, if one is not careful, can cause trouble.

One season, after my final undergraduate year at the University of Michigan, I worked for the Canadian Pacific Railway on its tie preserves in the Selkirk and Purcell Mountains in British Columbia, an unbroken wilderness. (About fifty years later, my wife and I drove over the magnificent Trans-Canada Highway across Rogers Pass in that general area, marvelling at the engineering feat that had accomplished this partial conquest of rugged terrain.)

On this crew I was classified as one of the more experienced men, working with a somewhat older man who had been out in the Alaska gold rush. Together we were breaking in two inexperienced college boys, running compass lines, chaining, and making a map and a cruise of the timber. One day we were in a burned area that had partly seeded up with scattered lodgepole trees growing in the open with a lot of branches. Old Hank Pottinger and I were in the lead, with the two boys waiting behind to stub the eight-rod tape when we got to the end of it. Coming up over a little knoll in this semiopen country, I noticed a bear a short way ahead across a gulch and remarked to Hank, "There's a bear." Of course, both of us had seen a great many bears, but Hank said, "Call the boys up. I don't suppose they've ever seen one." So I turned and shouted to the fellows, "Come up. There's a bear up here."

That was the first the bear had seen of us, and it was the first I realized that the bear had a cub. It was just a black bear, but she was startled. She whirled and charged in defense of her cub. Hank went up one tree, and I went up another. It didn't occur to either of us to tell the boys to stay where they were. They came bounding up over the hill. One of them was an all-American football center under old "Hurry-up" Yost at the University of Michigan, a wonderfully fine fellow. But, reared on German folk tales of the bear that ate naughty children, he really thought bears were very dangerous. Up he came over the hill. The other boy, Heath, reacted instantly when he saw the bear, whirled, picked out a tree, and went up it like a squirrel.

But big Hank Almendinger, who was built just like a bear and probably weighed about as much as this one, stopped in midstride, poised with his arms outstretched. Whereupon the bear, shocked at seeing a challenger, followed Hank's example and stopped too, fortunately. I doubt if they were over a hundred feet apart at the time. Then the bear rose up on her hind legs to study this situation. Well, of course that meant the war was over. When a bear stops to think it over, that's usually the end of the battle. She turned back to see if her cub was all right, then cuffed it and chased it up the hill. But I shall always remember that big football player, standing there ready to receive the charge, and I've always had a suspicion that he probably would have been the equal of the bear.

Of course, wild animals may be unpredictable. I said black bears are almost always harmless—but not always. One should use a modicum of common sense and precaution in dealing with all wild animals. I recall one fall, just after World War I, being on a crew making a survey to settle a large grazing appeal case near the northwest boundary of Yellowstone Park, at the head of the West Gallatin River. The chief of the party was Leon Hurtt, a University of Nebraska man, a capable technical grazing man, an excellent woodsman, who had a lot of practical experience as a rancher. Like all the rest of us he was out running lines.

One night when we were all seated at supper, I made some reference to having seen a bear that day. Whereupon Leon said, "Well, I guess I might as well tell you about mine." He spread his legs apart showing the inside of each pant leg. His trousers were frayed and completely worn through—those brand new wool trousers.

He told us he had stopped to write up his notes, then had decided that it was so close to noon that he'd eat his lunch first. When he got ready to leave, he suddenly noticed a black bear approaching. As we frequently did, he shouted to see the bear run. But instead of running away, the bear proceeded to charge him. Leon had to climb the closest lodgepole pine, which, unfortunately, didn't have lower limbs, so he had to shinny up it as best he could. The bear was a male, not a female with cubs. It kept him up the tree for what seemed an endless time. Since there were no limbs within reach, Leon had to clasp the trunk with his arms and legs and was continually sliding down and climbing up again. The rough bark simply wore out his trousers. Finally, the bear tired of the game and left.

Leon commented that he thought at the time that in all probability the bear was a park bear that had been bumming food from tourists. Maybe, he thought, if he were to go down the tree, the bear would go away. It's one thing to be convinced in theory and it's another thing to go down the tree and face a bear. Leon did just what I would have done—he stayed up the tree.

It is surprising to me that there are not more cases of people harmed by wild animals. People who would be afraid of a pet cow in a farm will fearlessly feed a park bear or trifle around with other wild animals in the national parks. After all, they should remember that they are wild animals. Several years ago, in Jasper National Park in Canada, I observed such a case. A car of well-dressed people stopped at a roadside to look at and photograph three buck deer standing at the timber's edge. This was in September, which was approaching rutting season for all the deer species. The bucks' horns were already polished; their necks were swollen. Buck deer who have lost their fear of people can be exceedingly dangerous at that time—dangerous in the same sense that a Jersey bull is dangerous.

One of the women started feeding cookies to one of the bucks, then began to tease him, offering a cookie, then jerking it away. As the buck became impatient and started stomping his feet, I warned her to get away. Those sharp front hooves cut like a knife and the antlers take their toll. There have been cases of animals of the deer family killing people—who have usually "asked for it."

In spring, deer present a more gentle picture, such as we observed one June. We were camped at Riverside Park on the South Fork of the Flathead River, in the area where—I'm sorry to say—the water from Hungry Horse Dam is now probably fifty feet deep. It was a beautiful, timbered wilderness area that has been ruined to create a reservoir. The deer loved to come out into that meadow, perhaps eight or ten acres in size, beside the South Fork. At that time of the year, deer readily become tame as soon as they know you are not going to harm them. In the evening, they used to like to get up in the smoke of our campfire where mosquitoes were less bothersome. Deer suffer from mosquitoes just as much as people do.

There was one little yearling doe that often got within eight or ten feet of us. It is an impression—and I believe generally correct—that bucks don't mix much with the other deer except in rutting season in fall. However, there are exceptions. In this group of deer, there was one old flop-eared buck with his horns just in velvet and only partially grown. One ear was damaged so that it flopped up and down—it was completely out of control. He seemed to be a good natured old chap. There were also several little spotted fawns—couldn't have been over a month old. The old buck used to play with one of them. He would chase the fawn around little cinquefoil bushes in the park. They'd go darting around and then stand and look at each other and pant and catch their breaths. Then in turn the buck would run and the fawn would chase him. This provided us with real entertainment.

Quite in contrast with this gentle scene is an experience I had some years later with another member of the deer family. In the fall of 1919, after I was discharged from the army, I was in a party cruising timber up on Big Creek on the North Fork of the Flathead River in Montana. It was one of those fall timber cruises where we were trying hard to finish the job and get things in shape so a timber sale could be advertised. We were racing the weather all of the time and living out of tin cans.

We had a typical old woods cook who slept in the cook tent where food supplies were kept. We'd had a little bear pestering around camp, eating our garbage. One night when the bear hadn't found enough to satisfy him, he went into the cook tent and stole a slab of bacon. We heard a big racket and popped out of our tents to see what was happening. There in the moonlight over the light snow, the old cook, clad in his long johns, was brandishing a piece of cordwood as he chased the bear. We had to laugh, but the joke was on us. Next morning he quit, saying he wouldn't stay in any camp where the bears came into the tent with him; and there we were without a cook.

So it became a case of having to do our own cooking. Jim Brooks, chief of the party, gave me the assignment first. I think I was chosen partly because my health was poor after the years in World War I, and he could see it was difficult for me to stand up to the fieldwork of running line in those cold, fall conditions. I was assembling supplies preliminary to getting dinner one afternoon, when I heard a noise outside the cook tent. I stuck my head out and saw a bull moose. He was a big one with a fine head. I noticed he had some fresh wounds. I think he was an old bull that had been chased out of the herd by some younger bull, and his disposition had simply gone to pieces. He was out looking for trouble, and I represented trouble.

I had a rifle there and could have killed him, but it was closed season on moose. I knew perfectly well that nobody would believe this forest officer had killed a moose in self defense. They'd think I simply wanted to kill myself a moose. So I climbed a convenient open-grown lodgepole tree with a lot of branches; it wasn't much more difficult than going up a ladder. I took the rifle up with me. I climbed it just in time before the bull charged. He went round and round the tree, stopping to lower his antlers and throw up pine needles mixed with snow.

It was tiring and cold, sitting up in the tree, so I picked off cones and tossed them over onto a tent nearby, hoping the bull would take offense at the tent and tear it up. Then I'd have tangible evidence to justify my killing him. However, he had no illusions; that didn't work, so I just sat still. Presently he went down the trail into the timber, and I promptly got out of the tree. Whereupon just as promptly he came charging back. He had gone into the timber just to deceive me and to peek around the brush. Well, that time I stayed up in the tree. Even after he left again, I stayed up there quite a while—it seemed an age.

Bull moose in fall are not to be trifled with. This old bull, on the warpath, met our packer on the trail as he went away and chased the packer and the whole pack string up the side of the mountain.

Getting down to modern times, years later we were fascinated with the adventure of a pair of chipmunks at Big Prairie Ranger Station in the Bob Marshall Wilderness. I had flown in with Bob and Dick Johnson of the Johnson Flying Service one day to bring supplies to the station, including oats for the horses. During the unloading operation, the two chipmunks got aboard the plane to nibble some scattered oats and were locked in when we made an extra hop downriver and back. When Bob opened the door on our return to Big Prairie, they skittered out, much to our surprise. Wonder what they had to tell the other chipmunks about the ride they hitchhiked? The first chipmunk astronauts!

Mountain lion on the prowl. —Photo by K. Fink, National Park Service

THE MOUNTAIN LION SCREAMS

There were assignments on which I was alone for weeks at a time, but the possible monotony and loneliness were mitigated considerably by my horse. A horse is often a lot of company. I had an exceptional horse in the summer of 1912, an Indian pony that probably, as a horse, wasn't worth over twenty-five dollars. However, he was the best camp horse I ever had. Buck was a mustang that had been raised by the Salish Indians, who, at that time, were largely blanket Indians who moved around a lot on the reservation. He wasn't, like many horses, tied down to any particular location as his home. Home with him was camp; that was the way he was raised.

Like many Indian horses, he had been raised very gentle. It was the white men who used to break horses by breaking their spirit, by spurring them and riding them until they feared all human beings and were ready to fight anyone. Buck was gentle. When we camped at night, I had to picket my packhorse; I couldn't trust that mare; she'd run away any time. But I merely took the saddle and bridle off Buck and turned him loose. I never had him leave me.

In good weather, I usually wouldn't pitch a tent but would merely roll my blankets out on the ground. Buck dearly loved to come around before he settled down for the night, nuzzle his nose in my face and smell me over, then lie down himself. He'd wake me up around two o'clock in the morning when he started to graze, but that was a minor disturbance.

Buck was scared of mountain lions, as was any sensible horse, for mountain lions are very fond of horse meat. At one place where we had a cabin with a little hillside pasture, there was also a mountain lion that used to scream, a terrifying sound. One evening, just at twilight, this mountain lion let out a terrific screech. Down came Buck from the hill, through the pasture, and over the fence. He never paid much attention to fences; he just took to the air and went over them. The first thing I knew, his head and front feet were in the door of the cabin. He was looking for protection.

Mountain lions are animals that are seldom seen because they are so clever at staying out of sight of men. They are not usually regarded as more dangerous to adult human beings, whom they fear, than pet kitchen tabby cats. One may, accidentally, see a lion out in the woods, but it is rare. I have known many men who have spent the greater portion of their lives in country where there were many lions but have never seen one—except, perhaps, one that had been chased up a tree by dogs or had been trapped and killed.

Nevertheless, they have a peculiar habit of following a man. Many a hunter, doubling back on his trail, has found mountain lion tracks following him. It has happened to me a number of times. I have stopped at some strategic point, when there was a little opening behind me, thinking I might possibly see the lion following me. I thought I might possibly ambush him, but it never worked.

In all my time in the woods I have seen very briefly only three mountain lions. One was the one that frightened Buck, or I presume it was the same one. It was in the same vicinity. It was early fall, and I wanted to get some venison to add to my meager rations. However, I didn't have a rifle, just a six-shooter, so the chances were dim that I could get anything unless I could get very close to some deer. Hopefully, I went down to a little marshy, sedgy meadow to see if I could creep up on a deer. After crawling quite a distance through the grass, I stuck up my head to see if I were going in the right direction. I looked squarely at a mountain lion who was doing exactly the same thing. He, too, was sneaking up on those deer. However, I didn't get to see him long. There was just a yellow streak and·he was gone, safe in the timber.

One night on the North Fork of the Coeur d'Alene River in Idaho, after dark, I was walking along a trail in dense hemlock timber with no underbrush, when I heard a little thud off to my right and could see two eyes. Occasionally, as I went through that bottom, I could see by the faint starlight, outline enough to know it was a mountain lion. There, again, was the curiosity of the beast, his apparent interest in people that will induce him to follow men. He wouldn't have gotten that close to me if it hadn't been dark. And, as soon as I came out into an opening where it was lighter, I saw no more of the lion. He constituted no threat.

The third time I saw a lion, my wife and I were sleeping in our car for a night on the Fisher River in the Kootenai drainage. In the morning, just at daybreak, a mountain lion walked right through camp. Apparently he could not catch our scent and walked into camp without even knowing we were there. That time it was a big fellow. I jumped out with my rifle, hoping to get a shot, but mountain lions don't stick around. The minute the car door made a slight click, there was that streak of yellow and the lion was gone.

That totals the three I have seen in the woods. Two or three times I have heard them, and I have often seen their tracks or where they have killed a deer.

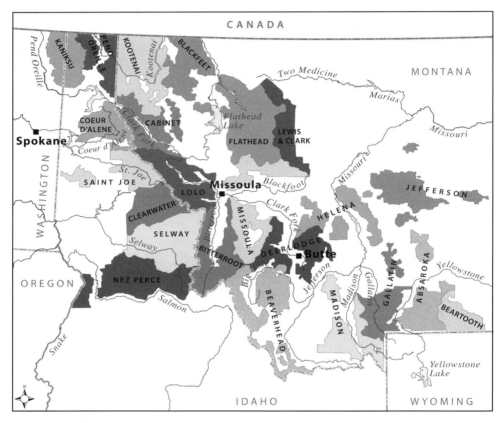

The national forests as they existed in 1913 in District One, the predecessor of Region One.

"No Cut Trees"

The Forest Service is firmly committed to the multiple use concept—to using National Forest lands in every way that can yield benefit to the community or to the nation. The lands are not exclusively for the raising of timber, for the grazing of livestock, for the propagation of wild animals, for recreation, or for the preservation of watersheds, but for all of these. But my initial concern was with timber. And, like any project connected with federalization, there was controversy over timber.

We had a hot dispute with so-called settlers up on the Little North Fork of the Coeur d'Alene River of Idaho in 1910. Here, alleged "homesteaders" had filed claims on some of the finest white pine timber in the world—filed under the Homestead Act. Many of the claims were not filed until after the area had been withdrawn from the public lands and incorporated into the national forests. The registrar at the Land Office allegedly had been bribed to back-date the applications to make them appear as if the filings had been made before the land was withdrawn and therefore were legal.

If the filings had been valid, and if the homesteaders had fulfilled the requirements of residence and land clearing, the claims could have been sold for fifteen or twenty thousand dollars each—a lot of money now, much more then. The lumberjacks who filed were mostly employees of a Coeur d'Alene lumber company. The registrar of the Land Office and official of the company were later convicted in federal court.

I was just a twenty-one-year-old boy, a temporary employee of the Forest Service, living in a little one-room cabin called Honeysuckle Ranger Station. Infested with packrats, it was a horrible place. I remember that one of the packrats there stole all of my prunes and replaced every prune with dry horse manure. It wasn't a pleasant place in which to live.

I came back to my cabin one night to find that the packer had been through and had left my mail. There was a letter from the supervisor, Roscoe Haines. He reported that the Forest Service had gotten a decision

A timber survey camp amidst the dense growth of the
Coeur d'Alene National Forest in 1915.

against one of the claimants by the name of Mumsford. He instructed
me to go up and tell Mumsford about it and stop his cutting timber. Now
Mumsford was a great big mountaineer who had a reputation of having killed two men and gotten off both times on a plea of self-defense. I
thought the thing over and decided that for $60 a month on this summer
job, it wasn't worthwhile facing Mumsford. We had all figured it was
going to take a flock of deputy marshals to get him and that some of them
were going to get killed when they moved Mumsford out of there.

However, in the morning I felt more courageous and decided I was
going to take a whirl at it. So I headed up toward the claim, but the closer
I got the scareder I got, too. As I thought how to approach Mumsford, I
decided that my gun (and well dressed gentlemen in those days wore a
gun) wouldn't be of any value to me in dealing with him. So I took off my
six-shooter with the holster and hid it behind a stump. I was not wearing
a coat, and I rolled up my sleeves so he could see I was unarmed. Then
I continued on to the claim. There was Mumsford bucking* a big white

---

* Bucking a log is when a man, working along with a saw, reduces downed tree trunks into
manageable lengths.

pine tree, sawing away, but when he saw me he stopped and picked up a Winchester rifle.

I continued to walk toward him until I was on one side of the log and he on the other. I tried to chat with him and extend the time of day and discuss the weather, but he wasn't interested. All he would do was grunt. Finally, I had to open the supervisor's letter and suggest that he read it. He took quite a while reading it and then read the last sentence aloud: "You will see Mr. Mumsford and inform him of this decision and stop him from cutting timber on his claim." He handed the letter back to me, dropped his hand back down on his rifle and said, "Well, kid, how are ya goin' to stop me?" I thought that over for what seemed a long time—it probably wasn't—and then told him the truth: "I'll be damned if I know."

Mumsford evidently did have a sense of humor. He had to laugh. Then he put the butt of his gun down on the ground, put his foot up on the log, and started to talk.

"Kid, I think we're licked; you know those easterners have got their minds made up. They're going to have these forest preserves. We could chase you fellows out of here all right and we could chase some more

A timber survey party in 1915 on the Little North Fork of the Coeur d'Alene River. From left, Josh Cope (a 1914 Yale graduate who later taught forestry at Cornell University), unknown, Ed Coulson (graduated from University of Michigan School of Forestry in 1912), and Dick Hamilton.

out, but they would send the army if they had to. We're licked. I'm going to pull out."

And he pulled out, and I never have known for sure if it was I who stopped him from cutting timber on his claim.

Maybe the concept of conservation was beginning to take root, though slowly. At one poor little homestead in another area, I saw a sign nailed up on a rail fence: "No Cut Trees." At least that homesteader was trying to conserve what resources he had.

Those were rough, rugged days, and they called for similar responses. Bill Hall told of how he was hired by E. A. Sherman around 1907 to take charge of the area south of Anaconda, Montana, in the Big Hole and a great portion of what is now the Deerlodge National Forest. Mr. Sherman, in his sympathetic and kindly way, explained Hall's duties. He also stressed that there had been wholesale logging of mining timbers, convertor poles, and of cordwood for the copper ore calcining pits* in Anaconda and for the Butte mines without any permit from the Forest Service. Hall had to go up there to settle that trespass, find out how much had been cut, stop the cutting, and initiate action to collect the value of the timber that had been cut. Then Bill Hall logically enough asked:

"Now Mr. Sherman, what authority do I have?"

To which Mr. Sherman replied, "Well, Mr. Hall, haven't you a gun?"

That was the way it was operated. Bill Hall went up there to size up the situation and presently managed to get some funds with which to hire men to estimate the amount that had been cut. Men called scalers could estimate the board feet in a log by measuring the diameter of the stump left when the tree was felled. However, it was necessary to work them in shifts. Certain of the scalers sat around on prominent points with Winchester rifles while the others made the estimates.

However, I would like to come to the defense of the men who were doing the cutting, to extenuate their actions somewhat. If the Butte mines and the Anaconda smelter were going to operate, they had to have timber. Until the forest preserves became national forests dedicated to use instead of mere preservation, there was no effective procedure, under the Department of the Interior, for the sale and cutting of timber. For three hundred years, public lands had been open to exploitation, and a president's order closing these lands couldn't change public attitudes this abruptly. This was a phase in the development of the country. If the men who operated these mines hadn't been aggressive individuals who didn't hesitate to

---

*The rock was heated to reduce the ore and drive off gases.

Cordwood cutter in French Gulch, on what is now the Deerlodge National Forest. Much of the cordwood used in the calcining process at the Butte mines was transported this way to roads or flumes in the early 1900s.

steal timber if necessary, we wouldn't have had that great mineral development, and Montana's advancement would have been slowed. They didn't think of themselves as doing anything wrong. Their code hadn't developed to one comparable to that of the present day.

Take old Nels Pierson, for example. He was known as the High Swede, though he stood only five feet five or five feet six; but there was considerably more man concentrated in that five and one-half feet than in most men a foot taller. I knew him personally and found him perfectly capable of taking care of himself in any situation. It is told that one time two drunken lumberjacks made a pass at him with double-bitted axes. Nels Pierson had a picaroon* with which he eliminated one of the lumberjacks, while the other one was discreet enough to retire. They had taken on a bigger army in the High Swede than they could handle.

---

*An axe-length tool with hooked head for handling small timbers.

But while Nels ruled with a firm hand, he respected those who held him to account, forcing him to comply with government regulations. He challenged me when I was supervisor of the Deerlodge National Forest to see whether or not he could bluff me. He had delayed payments on timber sales, cut across lines, and disregarded all letters. I suppose I could have gone to the federal courts to initiate action. This would have cost a great deal and taken time, and besides, Pierson recognized direct action best. So, with Ranger Ed Kurtz, one of the excellent old practical rangers, I went up with a log chain. Together we chained one of Pierson's cars, loaded with stulls*, to the railroad track and posted government property notices on the timber. After all, the timber hadn't been paid for and was still government property, so our action was perfectly legal. Nels promptly settled and wasn't the least offended. These were tactics he understood and could respect.

Years later, in 1936, I was transferred to the regional office in Milwaukee in charge of personnel management. Even though my forestry training had been in the Midwest, I found this a very different environment with forestry problems quite different from those in our western regions. Shortly after arriving there, I visited a timber sale on the Nicolet National Forest in Wisconsin. It was a small sale of birch. The cordwood billets cut on this sale were taken to a little factory where they made such things as tongue depressors, spools, and even toothpicks. Now I had come from the big logging country, which made me hesitate to write back to my former associates telling them that I had come down to the level of taking part in toothpick sales. Nevertheless, that illustrates the complete change from the extensive and often wasteful logging practices of the virgin woods to the intensive utilization that comes with denser population and more industrialization.

That was one of the changes to which we had to adjust. Of course, the adjustment didn't affect just me. My wife had her problems, too, but she adjusted more readily than I. One time she was with me in Washington, DC, for several months during which she noted a growing list of people who seemed to have a false picture of our western ways. When a beauty parlor operator topped them all by asking if there was "still all that shooting out west" that she had seen in the movies, my wife answered promptly, "Oh no! My husband hasn't shot a man in the last seven years." Great silence ensued. However, we adjusted, made many new friends, and have fond memories of our midwestern and eastern years with the Forest Service.

---

*Timber used as props in mines.

WE HAD OUR FUN

We westerners had a great deal of fun out of the fact that our easterner coworkers often got lost rather easily. In fact, we had one party of four men lost for three days. They had their packs along, fortunately, and finally got back to camp in good shape, just a bit hungry as they weren't skilled yet in hunting grouse or catching fish.

One spring, a young graduate of an eastern forestry school, Haasis by name, joined a party mapping and estimating timber on the Little North Fork of the Coeur d'Alene River in Idaho. His first assignment was to run a pacing and compass traverse up a sharp ridge that divided the main North Fork valley from its tributary, Iron Gulch, at the mouth of which the party was camped. Since the ridge had long before been burned over, and there was no merchantable timber, he was not accompanied by a cruiser but worked alone.

Finally, late in the afternoon, he decided it was time to quit and return to camp. He put away his compass and mapping equipment and took stock of the situation. By this time, he was far up the ridge; there had been several spur ridges, and he realized he did not know where camp lay nor how to get there. It did not occur to him that he could use the map he had made and with the help of his compass retrace his steps to camp. Instead, he considered all the instructions he had ever heard of how a man, lost in the wilderness, could extricate himself. There was one universal warning; he must not get to going in circles. So he took out his compass and, deciding that the shortest way to civilization (about thirty or forty miles) lay to the west, started in that direction.

Meantime, in camp, the party had finished dinner, and since Haasis had not returned, it was decided to dispatch search parties. One went up the North Fork and the other up Iron Gulch. It was reasonably certain that one of the parties would intercept him. But Haasis had taken another precaution. In hope that someone might hear him and report the direction of his travels, he announced himself at intervals. Accordingly, the North Fork

45

search party presently heard a forlorn cry: "Yo Ho! This is Haasis, going west." Thereafter he was known as Haasis Going West.

Let me admit, however, that I am in no position to ridicule Haasis. Since then, I have been hopelessly and equally inexcusably lost in many of the large cities of the nation where Haasis would have proceeded with confidence.

People vary in the degree of their interest in Nature. Some are in the forest working just for the paycheck; others are interested in the plant and animal life, and in everything they see. The latter were in the majority. We had one chap, however, who was there through a little nepotism; his father was an influential man. Plainly, he was uninterested in the job. He always dragged along behind, which offended us, because when we got headed for camp, hungry, we wanted to go. Yet if we went too fast, we left him behind. That meant we had to wait on the trail at intervals until he could catch up. Otherwise, experience had taught us, we'd have to hunt for him later.

We were coming down the trail toward camp one evening, stepping right along, with him about fifty yards behind, as usual. We were all annoyed at having to slow down for him. A big yellow jacket hornets' nest happened to hang on a tree near the trail—a tempting target. As we

A pack train arriving in a camp on the Little North Fork of the Coeur d'Alene River in the summer of 1915.

went by, one fellow gave the hornets' nest a good rap with his calipers and knocked it to pieces. The yellow jackets had just time enough to get out and become active when our friend came along, and they stimulated him to travel considerably faster. That was the objective.

We had another boy who wouldn't wake up in the morning. He was always the last out to breakfast. One morning the packer, Waldbillig, had just been out rounding up his pack string for a trip out to Ovando. He came in as we were at breakfast. When he happened to see our sleepy fellow in bed in his tent, he paused long enough to dangle his lariat around the boy's feet and his blankets, took a dally around the saddle horn, and dragged him peacefully down to the lake and into the water. Then he shook his rope free, spurred his horse, and rode on.

Inevitably, since the bulk of the men in the Forest Service at that time were young, there was a lot of this harmless horseplay. I think of Brad Jones (not the real name) and his companion and their antics in 1908 or 1909. They came back from several months in the Beartooth Mountains and Slough and Hell Roaring Creeks to the little town of Gardiner, on the north side of Yellowstone Park. After taking a few social drinks and feeling quite happy, they remembered that they had to take their horses to the livery stable.

As they rode down the street, Brad's companion took down his lariat, roped a barber pole, and dragged it down the street, just to show his pleasure at being in town. Of course, Brad had to emulate his example by taking his rope down and looking for something to rope. All he saw was a dignified, elderly school teacher in long skirts and many petticoats, so he roped her and down the street they went, one of them dragging the barber pole and the other leading the school teacher. Of course, that was a bit too rough to be overlooked entirely, and Brad was subjected to suitable discipline.

In 1912, I had passed the ranger examination and had a temporary appointment on the Kootenai National Forest near Libby, Montana. For a handyman assistant, the supervisor gave me a man named Braden, a school teacher who was a University of Chicago graduate and not a forester at all. When I met him in Libby, I found he had purchased the horse his employment agreement called for, a round old Indian pony, but he hadn't bought a saddle. He planned to ride bareback. I had three horses, my own, a saddle horse I had borrowed from the ranger on an adjacent district, and my packhorse. I put a packsaddle on his horse and packed his equipment and my supplies on the two pack animals. Then I put him on the horse I had borrowed, which belonged to the ranger's wife; a lady's riding horse, it was a gentle but spirited mare.

There wasn't even a wagon road out of the small town of Libby then. People came and went by railroad or mountain trail. The streets were paths with grass growing, and Braden's packhorse, which he was leading, decided to snatch a morsel or two. Braden, of course, was just hanging on to the lead rope paying it no attention, and when the packhorse put its head down to graze, the rope was jerked neatly under the mare's tail. Of course, the mare resented this, put a hump in her back, and before Braden knew what had happened, he found himself sitting comfortably in the middle of the street.

I dismounted, explained what had happened, quieted the mare down, and got Braden back aboard. We went two blocks before we had a rerun of the scene. Braden was becoming discouraged and, I think, never would have remounted except that two young ladies laughed at him sitting in the dust. This time, I tied his packhorse behind the one I was leading and helped him back onto the mare. However, by this time the mare had decided that this was a pretty easy proposition. She didn't have to carry that load if she didn't want to, so he had no more than hit the saddle than down went her head and up went her back, and she started to pitch again. It wasn't hard bucking; actually, I believe Braden could have ridden her if he had made up his mind to do so. However, he was getting pounded pretty hard, and he'd had about enough of this, so he decided he'd dismount. Well, dismounting from a bucking horse requires a great deal of skill if one is to land right side up—and he didn't have that skill. She was about ready to quit when she saw his foot come up, but then she gave one final jump and switched ends. It left him up in the air with no place to sit. Down he came. That was Braden's initiation into the Forest Service.

Finally, I got him to the ranger station. We started posting boundary, retracing old section lines and putting up boundary signs at regular intervals. He could run compass well enough to follow the blazed section lines; he could nail the boundary signs and return along the section line to the point where we had started. However, I always had to meet him in the late evening to take him back to the station. This was entirely unnecessary. If he got back to his horse, mounted, and turned the reins loose, the horse would come to the cabin. We were feeding oats there. I explained this to him, but he said, "No, Mr. Taylor, that just doesn't make sense. If an intelligent man doesn't know where he's at, a dumb brute doesn't." I told him that I thought that was sound reasoning except that in this case there was some question about the intelligence of the man. He refused to be insulted and still wanted me to meet him.

However, one time we were up a drainage where I had to cross a considerable ridge. I decided I could get my horse through, so I told Braden that I wasn't going to come back to meet him. He was going to have to retrace his steps to the station when he finished his line, and if he couldn't find his way, his horse would. I went on my way and put in a rather hard day of working my horse over the ridge, along the boundary, and down into the other drainage, and finally back to the ranger station. Braden wasn't there when I returned rather late. I cooked and ate dinner, and still no Braden. Then it got dark, and I began to feel conscience stricken about that poor boy, lost in the woods. I had just made up my mind to get my horse, ride up where he had been working, and shoot and yell until I found him, when I heard a hail from outside the cabin. A voice said, "Yo Ho!" I went to the door of the cabin and there came Braden, riding up on Baldy. He said, "I beg your pardon but can you tell me the way to Granite Creek Ranger Station?" I replied, "Just get off your horse, you're there." He didn't even recognize the station when he reached it.

We put his horse away and fed it; then I took him in and fed him his dinner while he told me what had occurred. Everything had been uneventful up to the time he finished his work and mounted Baldy to head back to the ranger station. He said, "I must have gotten confused some way. I couldn't find the proper trail and I rode and rode. Baldy kept wanting to go other ways. It began to get dark and I didn't know where I was at. So, finally, I just turned Baldy loose, and seemingly he knew where he was at all the time."

Ever since then when I succeed in finding my way around a city (which is the place where I always get lost), I say I was just like Braden's horse; seemingly, I knew where I was at all the time.

As I look back on those experiences now, I have a greater comprehension of and sympathy for those eastern young men struggling to adjust to a strange environment. Many funny and even tragic things happen to men who spend their lives working under frontier conditions—things that fill the kaleidoscope of memory. But the most terrifying adventure of my younger days was not tragic.

I was a western boy, raised in the West, and I had never been around a big city. My first experience with that was when I started back to the University of Michigan. I had been in the Cabinet Mountains alone much of the time all summer, working as late in September as I could in order to have enough money for school. Then I came to Missoula, disposed of my horse, and soon thereafter climbed on the train, headed for Ann Arbor. I planned to lay over a day in Chicago to see the city.

That was the most terrifying experience of my entire youth. I had been through the fires of 1910 and some other forest fires. I had traveled some of the wildest country in the United States. None of it compared with the shock of Chicago. When I got off the train there, I was completely bewildered. After two or three attempts to cross the street, and getting caught off-side every time and chased back by traffic, it occurred to me that if I did succeed in crossing, I wouldn't know what to do. Certainly, I was at least as unprepared for this strange environment as Braden had been for the mountains, and I didn't have a horse to rescue me. Finally, I solved my problem by taking a cab (a substitute for a horse) across to the central station and boarding the next train out to Ann Arbor. That was my first experience in a city. Subsequently, of course, I traveled many places, but I never did get to be a city man. The experience made me a lot more understanding of the trials of city men in the woods and humbled me.

East Meets West—and Welds

When I came out of the University of Michigan with my master's degree in 1914, having also passed a Civil Service examination, I received my initial appointment as a forest assistant, the entering grade for the professional service. There were equivalent titles in all scientific lines. The entrance salary was $1,100 a year, which amounted to $91.66 one month and $91.67 the next month, carefully balanced off so that by the end of the year one had received $1,100, perhaps minus one cent or so. It wouldn't have been plus one cent when the government was paying.

Except at his headquarters, the employee's expenses were taken care of, but if a horse were required, the employee had to furnish it. When I had been assigned as an assistant ranger two years earlier, I had to furnish two horses and, of course, my own bedroll and tent, but the rest of the equipment was furnished. Away from headquarters the government paid for the food, which wasn't an expensive item in those days. Food carried on a packhorse was not elaborate.

My initial appointment to the professional ranks was to a party mapping and cruising timber on the Deerlodge National Forest in the Big Hole southwest of Anaconda, Montana (I later became supervisor of that forest). Ironically, while I had never had any particulur interest in the engineering features of the job, I was assigned to mapping. Those whose primary interests were engineering were assigned to cruising timber. I believe this was deliberate as the Forest Service wanted to make us all-around men.

Jimmy Yule,* whose education was in engineering, learned a little botany on that trip, too. He came back to camp one day to report an unusual concentration of little Douglas-fir trees, six to eight inches tall,

---

*After an illustrious career with the Forest Service, Jim taught at the summer camp for Cornell University.

Timber survey outfit in the Big Hole country southwest of Anaconda on the Deerlodge National Forest in 1914.

Reconnaissance camp on LaMarche Creek in Big Hole country in October 1914. From left to right, Roy Brownlee, John B. Taylor, Richardson, and Ed Coulson.

on the south slopes of Mt. Haggin. He was puzzled, though, by their pink blossoms. It turned out to be a thrifty patch of pink heather in bloom— *Phyllodoce empetriformis*. He wasn't the first person to be confused by it. The linear leaves do somewhat resemble fir needles.

At that time, foresters were required to have much more training in surveying and mapping than is the case today. Now the organization is large enough to have specialists for that work. However, in 1906, when the Forest Service assumed control of 156 million acres of western forestlands, most of it was unmapped wilderness. Though we might theorize about the intensive European forestry methods, our first priority was to take an inventory of what we had. Moreover, the organization was not large enough to permit much specialization; most of us had to work at a variety of tasks. More intensive forestry and consequently more specialization came later.

In 1915, the Forest Service was pressing a suit for damages against the Great Northern Railway, alleging that the railroad had set fires in the summer of 1910 and had caused a large and destructive fire around Two Medicine Creek south of Glacier National Park and east of the continental divide. In April of that year, a small party of us was sent over there to run a survey control that was to be followed by timber cruisers and appraisers to estimate the extent of the damage. We went in from Kalispell, outfitting there and taking the evening train east to a small siding on the Great Northern Railway where our party would start out.

Charlie Farmer, Fred Mason, and I comprised the party. Late in the afternoon, after getting our outfit together, we remarked that we were going to the hotel to change into our field clothes. Jack Clack, who had a tendency to pull practical jokes on people, said:

"Why do you change into field clothes here? Why don't you change in the hotel at Lubec (the siding) and leave your clothes there: Then, if you have occasion to go down to Browning to a dance or anything, you'll have suitable clothing."

None of our party had ever been to Lubec, so we followed his suggestion and got on the train in city clothes, light underwear, silk socks, and oxfords. At 2:30 in the morning, the brakeman came through the car and took up our baggage checks. We didn't think anything of that, for it was often done at flag stops where there was no agent on duty. Shortly, the train stopped, we got off, and our field packs and outfit were rolled out of the baggage car door.

We found ourselves on the open, grassy east slope of the continental divide with the April wind blowing at fifty miles an hour, with snowbanks here and there, and not a building in sight—simply mountains and wind.

We got over on the lee side of the railroad grade where the wind was a little less severe, dug our field beds out of our packs, and crawled in, city clothes and all, not even chancing undressing in that temperature. It must have been well below freezing and by present-day figuring, the strong wind that always seems to be blowing in that pass must have produced a chill factor down in the twenty below zero range.

I awakened very early after a short sleep, dug my field clothes out of my pack, managed to dress properly in bed, and started out to find the Lubec Ranger Station. Of course, as Jack Clack had very well known, there was no hotel. Presently I located the station in a swampy place. It was not much of a station—it has long since been abandoned—and no one responded when I knocked. Knowing the ranger was a bachelor, I opened the door with my Forest Service standard key and entered, finding the ranger asleep in bed. I was thoroughly chilled by that time. On the window sill was a small individual bottle of Canadian Club whiskey, such as was served in dining cars for a single drink. The ranger was still sound asleep and I was very cold, so I helped myself to that whiskey and returned the bottle to its place. It seemed to me that I needed it under the circumstance. Then I waked up the ranger.

He was very courteous. We had never met, but I was from the regional office, so he was quite deferential. He built a fire in the kitchen stove, and then he remarked:

"Mr. Taylor, do you ever take a drink?"

"Well, on cold mornings I would consider it," I replied.

He continued, "Well, I have just enough to give each of us a half drink. It won't be enough for the whole group. Suppose we have it."

"That would be good," I agreed.

He went over to the window to get the bottle. Then he looked at me sharply. I tried to appear innocent, but apparently didn't succeed.

"I don't know you," he said, "but I'm getting acquainted fast."

The deference ceased from that moment. I have always remembered that morning at Lubec, and for a long time I tried to get a chance to repay Jack Clack for tricking us into going there so unprepared. I never succeeded.

Naturally, we westerners expected our associates from the east to become involved in humorous situations, and some of them did. Now those same men, gray-haired and full-fledged westerners themselves, chuckle at the stories, as, for example, does K. D. Swan, a fine forester and warm personal friend. He came out here around 1911 from Harvard University. I think he will agree that he was one of the tenderest of tenderfeet. He reported on the Lewis and Clark National Forest in the Belt

Mountains, where travel was entirely by saddle horse and camp outfits were transported either by freight wagons or by pack animals. Saddle horses were no longer in vogue around Boston and Cambridge, and K. D. had no experience with this kind of travel. Nevertheless, he met the challenge and quickly learned to keep up with the hard-riding local men and to cover forty or fifty miles a day on horseback. I suspect, however, that this meant some difficult adjustments and no little personal discomfort. He, himself, tells of the time he saddled the wrong horse and wound up sitting in the barnyard muck. But afterward, he saddled another horse and completed the day's work. Moreover, he went on to become a competent woodsman whose photographs of virtually unexplored wilderness areas became nationally known.

The most interesting and fascinating thing to me in the history and development of the Forest Service was not the progress, nor was it these stories of the trials and tribulations of some of its men. Rather, it was the confrontation of these two groups of men of radically different backgrounds with each other, and of their blending into one homogeneous body sharing the best traits of each of the two original groups. On the one hand were a lot of rather impractical, technically trained men, highly educated and converted into crusaders by Pinchot and his associates. On the other side were the practical western men. They were highly selected men of diverse education, often not more than common school or high school at best. All knew well the customs, prejudices, and problems of their locality, and all were experts in the practical skills necessary in this country.

Among them were men like that early appointee of Theodore Roosevelt, Fred Herrig; there was E. A. Sherman, who was chief inspector of this First Inspection District; and there was Nathaniel E. "Than" Wilkerson, who built the first cabin in what is now the Selway-Bitterroot Wilderness.

The important thing was that in spite of some joking back and forth, each group respected the education or the practical skills and local knowledge of the other. There was little jealousy. Generally, each tried hard to learn from the other. And each was willing and even eager to help the other. I think it is rare that two such groups of people have intermingled with equal harmony. There were inevitably exceptions, individuals who were intolerant and who were weeded out. The important thing is that they were exceptions and not the rule, that the distinction between the groups vanished so quickly. I still wish some psychologist would explain how the Glen Smiths and the Karl Woodwards came to understand each other and to merge and blend until distinctions were erased and no one thought of one as being the "practical" man and the other as the "college forester."

Of course, there were some mistakes, some merely amusing, others more serious. I have already told of such things as the young forester who accepted the loan of a pair of chaps, took off his trousers, and put on the chaps. He soon learned that wasn't the style of the day and corrected his error fast. Another, a Yale graduate, granted a free use permit on dead timber, which was permitted; but the "dead" timber turned out to be 100,000 feet of choice larch tie timber. In spite of all his training at Yale, he had evidently forgotten his dendrology and thought all conifers retained their needles all winter. This was winter, and the larch stand had shed all its needles and looked dead.

Karl Woodward, as I have mentioned earlier, was chief of our first survey party and a distinguished forester. Like many of the early foresters, he went into the education field, for new forestry schools were being organized and there was a strong demand for men with both technical education and practical experience to staff the faculties. In later years, he became dean of the forestry school at the University of New Hampshire at Durham. He was a decided individualist in many ways. He got his woods clothing from Abercrombie & Fitch, which didn't exactly resemble what most of us wore. Abercrombie & Fitch at that time had just imported the first of the spiral puttees, which had been adopted by the British Army in India. Karl chose to wear them regardless of what anybody said. He was generally known thereafter as Rag Legs.

Glen A. Smith was the chief of range management in this region, a big, jovial man who made friends everywhere. He was originally a cow puncher and his twenty-gallon hat was virtually a trademark with him. He was one of those practical men of the West who joined the Forest Service early. Incidentally, like all those who were successful, he studied. Anyone would have taken him for a highly educated man, and he was—but he didn't receive his education in a university. His wife, Cressie, too, was a western girl from a ranch near Fort Benton and a worthy partner for Glen.

One spring, Glen started out to make an inspection of the Nez Perce National Forest, which is down in the Salmon and Snake River areas of Idaho. There the season opens up much earlier than here in the high mountains. When he left here he was wearing heavy clothing suitable to Montana in April and May. When it turns hot in the Snake and Salmon River canyons, it is really hot. All of a sudden in spring, it goes up to 100 degrees or so in the bottoms of those canyons. Soon Glen's wife received a telegram that read "S.O.S.B.V.D.P.D.Q."* As usual, Cressie understood and sent him his light underwear and clothing at once.

Glen, at one time, was supervisor of the Custer National Forest at Miles City, Montana. There he became involved in grazing administration. He told of a time when, while he was riding on the Ashland Division, a severe storm came up. He drifted out of the higher hills west to the Northern Cheyenne Indian Reservation in lower country. Travel conditions were bad, making him miss a couple of meals. He came to a place where some hospitable Northern Cheyennes lived. They invited him in to eat, serving a luscious stew for which he was grateful. After he had eaten a big plate of it, they urged him to have more. The old Indian woman said, "Dig down deep; little puppy in bottom."

The majority of the men with little formal education, who were appointed through the old practical ranger examination, proved to be capable employees, devoted and deeply loyal to the Forest Service. Ed Kurtz was one of those whose devotion to his district actually shortened his life. In 1929, while I was supervisor of the Deerlodge National Forest, we had a very bad fire season all over Region One. Several of my younger men had been called to help with overhead on large fires farther west when a half dozen fires broke out on the Deerlodge. We were caught short handed. Knowing Ed's competency and dependability, I went to inspect his fire last, having been busy organizing manpower and getting control of other fires. When I finally got to Ed's fire, northwest of Anaconda in rather difficult country, I found he had been working night and day with scarcely any rest. He was then in his late fifties, and he was completely exhausted. Even so, he protested violently when I ordered him home to rest and put another man in charge of the fire. Ed never fully recovered. He continued on duty, but later that winter he entered the hospital. I shall always believe that whatever his condition was, it was due, in part at least, to utter exhaustion on that fire.

Ed knew his end was approaching, and he made one request. He asked that he be buried in his Forest Service uniform with his badge. Badges, naturally, were guarded; we didn't want them scattered around. Ed said, "You can manage some way to account for the loss of the badge." So Ed was buried in his uniform, wearing his badge. There were many others like him, but to me Ed stands with those at the top.

---

*BVDs were a brand of lightweight underwear of the time. PDQ means "pretty damn quick."

At times there was a failure of communication between the practical and the college men. At one time, my work involved trying to teach rangers to identify by correct names some of the common plants, so their reports would be more accurate and useable. There was one brush for which there seemed to be no generally accepted common name; therefore, I gave one ranger the Latin generic name, *Ceanothus*, which was generally used by botanists. A few days later as we were riding on range inspection, I pointed across to a brushy hillside and asked what the brush over there was. He answered somewhat impatiently, "Oh, it's that damn ferocious brush." That illustrates how we sometimes came out.

There were a few early men in the Forest Service who weren't above taking advantage of this difference in background, showing a little inferiority complex, perhaps, by sowing the other fellows' paths with nails whenever possible. Early in 1916, I was assigned as assistant to a ranger who was known to have almost an obsession for playing practical jokes on college men—jokes which were not always in good humor, to say the least. Some were in the annoyance class, but others assumed more serious proportions. He was irritated because he had to make an annual grazing report, for grazing was of minor importance on his district. He showed his resentment in a very petty way. After compiling the required information in plain language, he told me to put it into the most technical language possible. He said, "Use the longest words you can, and if there's anything that has a Latin name, use the Latin name."

Accordingly, I made the report as verbose and technical as possible. Illustrative of the general tone of the report was the answer to the question regarding the occurrence of prairie dogs, which on many ranges were a pest. The ranger wrote for my use, "Any damn fool knows there aren't any prairie dogs west of the main range." My translation was: "It is generally known that *Cynomys ludovicianus* does not occur west of the main cordillera." The report came back, of course, for revision. It was a deliberate attempt to embarrass the assistant supervisor, whom he disliked.

Another trick he pulled was much more serious. In those days, firewood was the main fuel in towns out West, and the only fuel at the ranger stations. One of the ranger's jobs in fall, for which he was allowed time, was to get in a supply of firewood. But that involved work—hard work—cutting firewood. The general public could get permits to cut dead wood for fuel, without charge. The permit was required merely to make possible some control and to prevent infringement on places where the timber had been sold. The permit requirement was not always rigidly enforced. Many people going after firewood went past the ranger station, which was on the main road west of Anaconda. When fall came, the ranger merely kept his eyes open for someone coming down from the forest with a load of wood. Then he went out and said quite officiously, "Well, let's see your permit." If the man had no permit, he said, "All right, pull right over here and unload it." By stopping others, he soon had his winter supply of wood. Of course, as soon as that became known at headquarters, he was checked on that; but it illustrates his methods. He was ultimately discharged.

He wasn't above pulling other tricks covering a broad field. He carried on a virtual vendetta with the technical men who were continually being assigned him to assist in the work and to acquire on-the-job training and practical experience necessary for advancement. When I was assigned to him, my papers showed I had graduated from the University of Michigan, but he didn't know I had quite a few summers of experience in the Forest Service or that I was a western man. The hazing started. When I arrived at the station, driving my little model T Ford roadster, he was watching for me and met me at the gate.

"I'm the district ranger here," he said briskly. "I take it you're my new assistant."

"Yes," I replied.

"Well ," he said, "let's have an understanding right now. I'm the boss. Anything I tell you to do, you do; and anything I don't tell you to do, you don't do. I run things here. Do you understand that?"

"Yes," I answered, "that's quite clear."

"Well," he went on, "put your jitney down in that shed. You live over in that bunkhouse. The bell will ring when it's time to eat." That was my briefing.

There was very little said at dinner that night; that is, very little said by him. His wife and mother were friendly and cordial. I returned to the bunkhouse. The next morning he woke me up in good season with the bell, and I went to breakfast. Then he had me help milk two cows, which was quite all right with me; I was raised on a ranch. He was quite surprised, however, to find I could milk. That was merely a test. After we had milked and done the chores, he announced that we were going to take some supplies up to the lookout. It had been agreed that I could rent a saddle horse from him, and he brought me an old saddle-marked mare. Knowing his reputation, I was rather suspicious of this mare. Sure enough, when I put the saddle on her, she put a hump in her back and gave all the evidences of trying me out. So I put on a pair of spurs. I never was good at riding mean horses, but with this old mare I felt confident that probably all she did wouldn't be serious.

The ranger said, "Oh, you wear spurs do you?"

"Well," I replied, "you got any objection?"

"No," he said, "scratch her all you want."

As soon as I mounted, she started bucking. It was not very violent, nothing I couldn't ride readily enough. As soon as I was well seated in the saddle, I took him at his word and used the spurs. She didn't care for that. She quit bucking.

Whereupon the ranger said, "Well, I guess she's all tired out. You can ride that white one in the corral."

"Do you want another one broke?" I commented.

"No, he doesn't do anything." And the white was a good saddle horse.

Subsequently, I found that several men who had never ridden had been thrown by that old mare. That was the ranger's idea of a joke. He took enjoyment in being brutal, it seemed to me. However, his attitude was uncommon. Very few of the western men wanted to take advantage of the eastern boys in any such fashion. Of course, after this experience, he asked me where I was raised and found out a little about my background. From then on, I never had any difficulty with him, but he never missed a chance to cause trouble for other college men, regardless of rank. For example, one day an inspector from the regional office was scheduled to look over one of the many timber sales on the district.

Some of the timber sales were not in good shape and wouldn't have reflected well on the ranger if subjected to careful inspection. Therefore, the ranger instructed me to make no comments whatever, regardless of where he went. One evening, the inspector arrived, a man from New England with a broad Harvard accent, against whom, largely because of that accent, the ranger had a strong dislike. He spent the night at the station. The next morning I was told to saddle up the horses. That surprised me, for we could reach every timber sale on the district quite readily by car, and I didn't see any need for horses. However, I made no comment. An interesting day unfolded. We three started out on horses, forded the stream, and rode up the mountain, a most extraordinary direction if we were going to the timber sales. We rode around here and there by various trails and finally dropped over the hill to a timber sale that was really in fine shape. But it wasn't the sale the inspector wanted to examine.

Not knowing the country, the inspector was very confused by this time and took the ranger's word for it that he was on the right sale. After a thorough inspection, we retraced our devious route back to the station. The inspector in due time made an excellent report on the wrong sale. However, in the long run, such actions did not help the record of the ranger, and, as I have said, he was discharged.

It was the college group that most quickly and thoroughly learned the new skills and attitudes the job required. This was no doubt due in part to the fact that acquirement of woodsmanship was essential to their survival. Those who didn't adjust soon gave up and resigned. The practical man faced no such imperative need to change. If he failed to study and grow, there were still jobs to which he could be assigned if he were ready to forgo advancement. However, there were few practical men who did not acquire at least some measure of proficiency in fairly technical tasks, and many devoted themselves to systematic study and became highly qualified technical workers.

Reconnaissance pack train moving camp in August 1920 on the Upper Taylors Fork of the West Gallatin River, in the Gallatin National Forest in Montana. —Photo by W. R. C. Jr., US Forest Service

Cattle on the Deerlodge National Forest. —Photo courtesy US Forest Service

RANGE WARS—REGULATION COMES

The idea of requiring permits to run stock on federal land came as a terrific shock to the ranchers. The thought not only of paying a fee, but of having a bunch of rangers telling them how to run their stock and where to place salt and limiting the numbers they could run, was offensive in the extreme to these highly individualistic ranchers. I can see their side of it very clearly. They developed this country, and their whole economy was based on the use of free public range. Nevertheless, times were changing. It was inevitable that sooner or later they were going to have to be regulated in their use of range, because there was increasing competition. Range wars that have been written up fully in Wild West stories resulted from this competition for the ranges. Of course, with the Forest Service managing the ranges, competition between ranchers ceased. The competition then became a battle with the Forest Service as to who would get a permit, how much stock would be permitted, how soon they could turn out their herds in spring, and when they had to get the stock off the range in fall. The Forest Service was inescapably the target of a lot of animosity from these stockmen.

My first real contact with range management came after World War I, in 1920, when I was assistant supervisor on the Gallatin National Forest. Even though we were under attack a great deal by the stockmen, we found them a wonderful group of people. When we really understood their problems and succeeded in doing a good job, things were fairly peaceful. Yet I doubt that we were ever accepted as peers by the stockmen, even those of us who were raised on ranches and were western men. We were government men and on the outside.

That doesn't mean that they were unfriendly. Even where there was a strong disagreement one could stop at a ranch and be sure of a cordial welcome. Ranchers were hospitable. Forest Service men were received hospitably and treated well. Even though one might be arguing bitterly with his host the next day, as long as we were his guests, we were guests—and that was that.

My experience in range management and allied grazing problems became more extensive after I was assigned to the Division of Range Management in the regional office of Region One. In 1922, I was sent on grazing inspection of several forests, starting in spring on the Nez Perce National Forest in Idaho. If you want scary riding, try Hells Canyon. This canyon of the Snake River, west of the Seven Devils Mountains, forms the boundary between Idaho and Oregon. It is considerably deeper than the Grand Canyon of the Colorado River, very much narrower, and more precipitous.

It is not as scenic, nor has it the coloring of the Grand Canyon, but it is an inaccessible area, penetrated only by trails that I cannot term other than terrifying—and I have ridden plenty of mountain country. It used to be a standing joke among the local forest officers there that those of us from other parts of the region spent most of our time leading our horses and walking. They rode those trails regularly, of course, and became accustomed to them.

I made a resolution that I was going to stay on my horse as long as the local men did on theirs. For two days I managed to do so, although I wasn't happy. We were on trails where I could look over the side of my horse and stare down two or three thousand feet. However, we finally came to a place that was too much for me. I was told that one could sit on his horse there and lightly toss a pebble off the right side and it would drop 1,500 feet before it struck. I was told that; I didn't look. And the trail was narrow. I dismounted carefully and tiptoed around the point, leading my horse. After the mountainside eased off a little I got back on my horse. My associates looked back at me and grinned.

We had gone but a short distance when we met a woman—on horseback, of course—carrying a baby in her arms and with a little girl behind her holding on to her mother. The trail was so narrow that it took considerable maneuvering to work our horses off the trail so she could pass. She was a local ranch woman. There are ranches down in the bottom of that canyon because of the excellent grass. People do live in those isolated places. She was probably riding to an adjacent ranch.

I was in a state of shock. After we had gone a little way, I asked, "How on earth is that woman going to get around the point and down to the river with those children?"

The ranger said, "Well, she isn't like you. She'll just sit on her horse and ride."

With that in mind, I stayed on my horse thereafter, but I was very glad I didn't have to make any more trips into Hells Canyon. But those rangers were not above taking advantage of my determination to ride. They didn't want to make things too easy for an inspector. One night we stopped beside the Salmon River. The next day was frosty and cold. When we started out, I somehow found myself in the lead. I didn't think anything of it—but soon I did. Presently, we came to a bridge across the Salmon, a swinging cable bridge, possibly six feet wide, used only for saddle horses and packhorses. I didn't like the looks of it, for it was up in the air perhaps 150 feet above that rushing stream. However, I rode out onto that frost-covered bridge. It started swaying. There wasn't anything then to do but to keep riding. I couldn't get off the horse there, so I sat firmly looking straight ahead between my horse's ears, with the bridge swaying in an ever-widening arc.

I couldn't hear the ranger, Bill Daisy, or the assistant supervisor, Clyde Fickes, coming behind me, but I didn't dare to turn around to look for them. When I finally reached the other side, I stopped and looked back. They were waiting for the bridge to stop swinging. Finally, Clyde came out on foot, gingerly leading his horse. After he got across and the bridge had calmed down again, Bill followed, also on foot. The bridge was under repair and the side guy wires had been taken off. Those associates of mine thought it quite a good joke to let me ride out on it and see what I would do under those conditions.

At the time I went to the Deerlodge National Forest in 1923 as supervisor, there was quite a controversy over the grazing rights going on there. Such disputes were common on the ranges, partly due to the fact that the ranchers were individualists who didn't welcome regulation by any agency (particularly a federal one), and sometimes because the Forest Service hadn't done a good job of regulation. We made our share of mistakes.

This specific controversy involved a range I had never seen high up in a mountain basin, comprising about 10,000 acres of what—if it had occurred 2,000 miles farther north—would have been called muskeg. At its altitude, the basin was essentially a boggy area of redtop and sedges, the vegetation of northern swamps. Early in the season, I arranged for a group of the complaining ranchers to accompany me to the range to show me their problem. After riding quite a few miles in the mountains through dense lodgepole timber, we emerged suddenly to a flat expanse of waving grass. For a moment my eyes did not adjust to the sudden change except that I did detect an animal out in the meadow. I spoke much too quickly, "Well, some steer got up here already."

Guy Stambaugh, manager of Deerlodge Valley Farms, and Bert Cole, ranger on the Deerlodge District, in the Cable Range around August 1934.

The moment I said it, I knew what a fool I was making of myself. One of the ranchers commented, "Yep, can't tell whether it's a shorthorn or a Hereford, but it's got a dewlap."* Clearly, it was a moose. A forest officer had to guard his tongue carefully to avoid making just such little mistakes as I had made. The stockmen were quick to seize on such an incident and to spread the word that the new supervisor didn't know enough to tell a moose from a steer. Moreover, all too often our men, especially the college men, but often the rangers from the lumber woods, actually were quite ignorant of the stock business. The Forest Service tended to think first of timber, and range management in the early days was often relegated to a minor position in importance.

As the years went on, universities having forestry schools offered courses in range management, a trend led by Charles Edwin Bessey at the University of Nebraska. Moreover, a high percentage of the men who followed this specialization were from midwestern farms or western ranches. This gradually built up a staff of competent range men who could

---

*A dewlap is the pendulous fold of flesh on the throat of a moose.

hold their own with the stockmen. In the meantime, difficulties led to friction and to some very justifiable complaints.

We had one young appointee who was released before the end of his probationary period because of lack of adaptability but who managed before he went to make possibly the most foolish and famous mistake one could have made. He stopped at a ranch and, of course, was made welcome. During the afternoon his host showed him a purebred bull he had just bought of which he was quite proud. The boy considered it thoughtfully, looking it all over, then said, "Yes, a very fine animal. What do you use it for?" As you can imagine, that joke went all over the adjacent country. Perhaps it was a public service in that it furnished entertainment over a wide area, but it did very little to enhance the prestige of the Forest Service that one of its men didn't know the purpose served by a bull on a stock ranch. We couldn't afford that type of mistake. He was transferred at once away from there—far away. Ultimately, as I have said, his service was terminated.

Gradually there came, though grudgingly, a recognition on the part of the stockmen of the necessity for some regulation. This did not prevent criticism of how we administered the job. I believe the attitude of the more intelligent stockmen, the leaders, was pretty well epitomized by an elderly man near Melrose, Montana, a Mr. O'Connor. I understood he had come into that country as a pioneer, driving an ox team. When I knew him, he was well along in years and was a first-class stockman.

East of Melrose was an area outside of and adjacent to the national forest. The Bureau of Land Management had not yet been established, nor had the Taylor Grazing Act been passed, so that tract was open public domain, available for the use of anybody who got there—first come, first served. As a result a blade of grass could hardly get two inches above ground before a sheep or steer came to bite it off; soil erosion was starting. The range was literally being eaten down to the grass roots. A group of ranchers conceived the idea of petitioning Congress to have it added to the national forest, and thus to place it under regulation. Remember, I had had bitter arguments with those same stockmen over the distribution of grazing privileges on range within the forest. Now they proposed that the Forest Service absorb this large area, at least two or three townships, of public domain—free range. They invited me to the meeting at which they discussed the proposal and finally asked me to comment.

I rose and told them how we would undertake the job, how we would survey the range, estimate its carrying capacity, and how we would study conditions. Then we would try to get the history of grazing on the tract and to find out who, by long use and ownership of adjacent ranch lands, was most entitled to permits.

"But," I continued, "you should all bear in mind that that range can be put into decent condition and its carrying capacity built back in only one way. That's by a drastic cut in the numbers of stock grazed and by enforcing much shorter seasons of use to give the forage a chance to recover. When I'm doing that, I suspect you'll all be calling me an unqualified S.O.B. Now it's your business as citizens to decide whether or not you want to petition Congress. I have nothing to say on that; I'm just your public servant. I merely want to tell you what will take place if that's put under the Forest Service." And I sat down.

Then Mr. O'Connor got up and said something like this: "Well, fellows, that range is in awful shape. We all know, too, that the Forest Service is the only outfit that can regulate and get it back in shape. I've called Forest Service men S.O.B.s a lot of times, and I'll likely do it again if I feel like it, but they're the ones that can handle it. I move we petition Congress."

The tract was not added to the forest but finally came under regulation through the Taylor Grazing Act of 1934. However, the incident pictures how the stockmen looked at it. We were still outsiders; they were still asserting their rights to criticize their public servants as they saw fit, truly an American citizen's prerogative.

Another man in the same locality, Ike Edinger, was much the same type as Mr. O'Connor. He didn't hesitate to tell us when he disagreed with us, yet he was a broadly tolerant man. Generally, stockmen were ready to challenge us for any failure to enforce regulations equally on everyone, but there were exceptions. Mr. Edinger recognized the exceptions. I and one of the rangers were riding with him over his sheep range one day when we came upon a half dozen Holstein cows right in the middle of his sheep range. I didn't know where they had come from; the ranger was not on his toes and didn't know either. Of course, I immediately started to quiz the ranger, whereupon Mr. Edinger broke in: "Don't ride the ranger too much about those cows; they belong to that widow who lives down in the mouth of the canyon. Her husband died last year and she's having a hard time. Let them run."

One can't talk about the range without being asked about the battles between the sheepmen and the cattlemen. When the range was out under administration, both groups had to apply for permits. This eliminated their main reason for fighting each other, and they turned to battling us. Both wanted a certain range. Both brought every possible pressure to bear on us to get it. When the issue was finally settled, I don't believe we got a

Drop band of Mt. Haggin Hampshires in spring 1927 on the Deerlodge National Forest. A drop band is a group of ewes about to have lambs.

great deal of gratitude from the man who got the range, but we did win considerable criticism from the one who didn't.

There was one class of people that was favored by the regulations. The government being ruled by the majority tends to favor the little man, and the small homesteader and ranchers were deliberately given an advantage in grazing privileges over the large outfits. Whether this was wise or not is debatable—those small units were not very efficient operations. Nevertheless, that's the way it was.

In spite of this favoring of the small operator, about eight out of ten of our complaints and appeals came from these same little foothill ranchers with thirty or forty head of cattle. Lots of times, I think the complaint was not based on what we had or had not done. These people lived out there isolated; here was one place they could make themselves felt. If they put in a good appeal and carried it on long enough, they might even be successful enough to get the Secretary of Agriculture to look at it. They could not be sure it would go that far, but they could be certain, sooner or later, of getting an official from fairly high up to study their complaint case. They got some attention and, consequently, a feeling of importance.

I remember a complaint on Limestone Creek, way up in the Beartooth Mountains, an incipient appeal from an isolated rancher who, with his wife, maintained a little home and had two or three milk cows and thirty or forty head of beef cattle running on forest range. When I got there in the afternoon, the man was away—out on the range—but his wife was insistent that he would want to see me and that I wait for his return. Time went on and evening came. She said, "Wait and have dinner with us. He might be a little late."

Meantime, she was getting out the pails to milk her two cows. Noticing that, I spoke up, "Well, if I'm going to take dinner with you, and you've got to get dinner, let me milk the cows." She looked at me with considerable surprise and asked, "Do you know how to milk?" I replied, "Of course, I was raised on a ranch." So I milked, and while I was still on the job, the husband came home. I saw him and his wife conferring in whispers with each other. We had a good dinner. Afterward they chatted with me in a friendly way and we had an enjoyable visit. Finally, I said: "It's a long ride back to the ranger station. Wonder if we can't get down to business, and then I can get started." He answered, "Well, I guess we'll drop our complaint. We don't want to fight about it with a guy like you."

It may sound strange, but I had become one of them, at least temporarily. I had milked their cows, we had had dinner together and a pleasant evening, and they were more ready to accept me. To the extent that local forest officers could get on that footing with the stockmen, things went well; but where, as was inevitable, we got in fellows who were not tactful and could not see the ranchers' problems, there was friction and trouble.

One of the big problems in the early days of the Forest Service in covering our districts was that of travel in rough country. Today, we have roads in every direction, plus telephones, radios, automobiles, airplanes, and helicopters, not to mention snowmobiles in winter, to enhance our communication picture. In contrast, we also have special areas designated as wilderness areas where we have to hold fast against pressure from people who wish to push roads in. Then, however, we had to travel by foot or horseback over a large part of the country.

We often used to find ourselves stopping with one of the trappers, hermits, or prospectors who lived back in the hills. I think you might roughly divide them into two classes. There were some who were meticulous in their housekeeping; others were quite the reverse. However, without exception, they were always friendly and would take us in for a night. In fact, they seemed glad to have company.

Government regulations required us to pay for meals and lodging, fifty cents or so for each, if we could possibly get our host to accept it. After a while, most of these people came to realize that that was a peculiarity of forest officers, that they were insistent on paying for meals and lodging. We assured them that we were reimbursed on expense accounts, so they put up with the practice, although they didn't like it. It wasn't the western practice to accept pay for meals and lodging in private homes. You were privileged to stop wherever you were and your host was not the least disturbed about it because sometime or other he expected to stop with you. That was the code of the West.

On the larger ranches, I have been places where the problem of paying was handled by the simple expedient of paying the cook. Men like Charlie Anceney of the Flying D Ranch simply would not take pay for meals or lodging or for riding one of their horses. However, we could comply with regulations and soothe our consciences, and Charlie Anceney could save his pride, by our paying the cook, who never objected.

The Deerlodge National Forest is one of the farthest west of those in Montana where grazing is a major activity. It was natural, perhaps, that one time the National Woolgrowers' Association held its annual convention in Butte. This was during a time when the association was financing a national advertising campaign to eat more lamb. We tried to work closely with the news media—reporters used to stop in my office regularly for news—and shortly after this convention opened, Ed Hileman, one of the reporters, dropped in. He said he simply had no news and that I had to dig him up some sort of story. I replied, "Everything's dead with us, too. I haven't any news. I have an idea, though, that might make a story. Why don't you go down to the convention headquarters in the Finlen Hotel and see what those woolgrowers are eating—whether they're eating more lamb or not."

Ed immediately got the idea and down he went to the Finlen Hotel where he cornered the head chef. He found that the Finlen had laid in a good supply of lamb to supply the woolgrowers. The chef said he had had, including the annual banquet, only three orders for lamb. All the sheepmen were eating beef. It made a wonderful story, but I had to put Ed under promise not to reveal where he got his suggestion.

Wild horses on open range in Montana. —Photo courtesy Bureau of Land Management

Wild Horse Roundup

On those national forests where grazing was important, a serious problem developed with wild horses. Some of these horses were descendants of the original mustangs, the feral horses that escaped from the Spanish in the Southwest and multiplied and scattered all over the West. It was from these mustangs that the Indians supplied themselves. Some of the tribes bred them up to very good strains of horses, notably the Nez Perce with their Appaloosas. Most of the horses on our ranges were merely escapees from local ranches and their progeny, a bunch of inbred, undersized knotheads, virtually valueless. There were also a few recent strays from ranches, good branded horses that their owners were anxious to recover.

For good reasons, the stockmen considered these stray horses a pest on both cattle and sheep ranges. Not only did they eat up a lot of forage, but on cattle ranges they would gather around salt grounds and water holes and chase the cattle away. To have maximum forage to enable the stockmen to prosper, the horses had to be eliminated. The roundup of those wild horses was a very unpopular job. Our rangers hated it. Though we hired excellent riders to staff the roundup, it was a job on which there were bound to be minor injuries. Occasionally a horse would fall, bringing bruises and possibly broken bones to its rider. The only way to round up wild horses was to get behind a band of them and to keep them running until they were worn out. Our men were riding good, grain-fed horses and could arrange for relays of horses. They could outlast the stray band.

The men one hired for those roundups were naturally young, adventurous riders. It took men who were willing to put their horses at a dead run through rough country behind a bunch of wild horses to do the job. The only way to get real enthusiasm on those roundups, however, was for the supervisor and ranger to get out and do a little riding too. I had to do that and am free to admit I didn't like it any better than my rangers did, but I joined several roundups. Following the custom in that area, the man in charge on any given range would usually be one of the local ranchers who then cooperated with us in ridding the range of horses.

On one roundup, the foreman was somewhat of a practical joker and also probably thought it would be a good idea to test the supervisor. At any rate, he assigned me my horse, which, to my surprise, I found to be a gentle one. I had expected him to hand me something that would give me plenty of trouble, but it was a quiet, gentle horse and strong.

We had gathered together a fairly large herd of wild horses that we wanted to take down the road between barbed wire fences three or four miles to a larger pasture. There we could hold them for sorting and final disposition. The foreman suggested that I ride point. That is, along with two or three other men, I would ride in front of the herd, with the wild horses at a dead run, of course, until we got them quieted down. The rest of the men followed behind, chasing the horses.

We hadn't gone over a quarter of a mile until my horse stumbled and started to fall down. Well, a rider can hold a horse up to a certain extent, and I certainly held that horse up—with about two hundred head of wild horses behind me. I managed to keep from going completely down; we didn't fall. Soon, however, we came to a fork of the road. There, one of the point riders would have to swing out to turn the herd to make sure it

John B. Taylor, supervisor of the Deerlodge National Forest.

didn't go the wrong way. I swung out and let the herd go by—with relief. Later I learned that my horse was known as a tangle-footed brute. That was the type of rough joke occasionally pulled by the reckless men who took part in these roundups.

We conducted quite a few of these roundups and ultimately got the range cleaned up. All of the branded horses were returned to their owners; the others were put up at auction. The better stock horses would be bought up at prices as high as fifteen or twenty dollars each. Finally, the remainder, the bulk of the herd, old horses, small mustangs, inbred worthless creatures, were sold to a packing plant where they were butchered. The better cuts of meat were pickled and sent to Europe for sale in Belgium and France where horse meat is valued. The remainder went into cat and dog food. Hides were sold and refuse made into fertilizer. The money collected went toward defraying the cost of the roundup.

We used to have many protests from people whose sensibilities were offended at the thought of killing horses. I remember two elderly ladies who came into our office to protest our selling about five hundred horses to Hansen Packing Plant. One of them said, "I think of my dear old carriage horse when I was a girl, and it just makes me cry." I assured her those animals were not to be compared with her fine horse. I wonder what her reaction would have been had she entered a corral with one of those wild horses. It would have streaked toward her like an infuriated wild animal. That's what they were, just wild animals. I never saw one of the noble, handsome stallions portrayed in fiction as leading and protecting his band. Frequently the wild band would be led by some wise old mare, and the stallions would be content to seek safety in the middle of the herd.

There is active controversy today between cattle and sheep men and those who oppose any action to dispose of wild horses. As usual, both sides take extreme positions. A reasonable compromise would be to designate a few ranges for wild horses and to remove them from all other ranges where they are injurious to the grazing of livestock. Even then, it would be necessary to reduce the horse herds from time to time or they would increase in numbers until overgrazing would result in winter suffering and starvation. There are not enough predators to keep the herds within the carrying capacity of their ranges.

*Starting game patrol in the morning from the Cinnamon Ranger Station
on the West Gallatin, Gallatin National Forest, November 1920.*

ELK HAD PROBLEMS TOO

My first active work in game management came shortly after World War I. I had, like most Forest Service officers, been a special deputy game warden to enforce the state game laws whenever we encountered game violations. However, in 1920, on areas adjoining Yellowstone National Park, the Forest Service, the State of Montana, and the National Park Service joined forces in a cooperative effort to protect the elk herds.

At that time, there was an active market for elk teeth, the two teeth from each elk—vestigial tusks—resembling ivory. They are often beautifully colored and were cherished by people for watch charms or cuff links. At that time, also, members of the Elks Lodge often used the elk tusks with their emblems. Later, realizing the depredations on the elk herds, the Elks Lodge outlawed them and cooperated in helping kill the flourishing market. A good pair of well-colored teeth from a bull elk had been known to sell for as high as fifty or seventy-five dollars. Cow teeth are small and were of little value. However, some of the men dealing in elk teeth used to put the cow elk's tusks into a solution of chewing tobacco and water. Left in this solution for about a month and then polished up, they became beautifully colored. It is true that in a few months the color tended to fade. Nevertheless, the process proved to be profitable. A valueless pair of teeth, if colored with five cents worth of Star chewing tobacco, might bring as much as ten dollars from some easy mark.

Around the northern side of Yellowstone National Park, a group of rough men used to engage each winter in tooth hunting. When the elk were driven out of the park by heavy snow to lower country, they tended to band up on what was then the Absaroka National Forest east of the Yellowstone River, and the Gallatin National Forest west of the river. In the fall of 1920, as assistant supervisor of the Gallatin, I was senior officer with a group of five or six men, all working together to try to protect the elk herd on the head of the Gallatin River drainage. Another group was similarly operating in the Tom Miner Basin area and north of Gardiner,

Montana, on the west side of the Yellowstone River, while a third party covered the country east of the Yellowstone and on Slough and Hell Roaring Creeks.

The people with whom we were dealing were a rough gang—a bunch of hillbillies who could not be trusted to confine their shooting to elk. Of course, when it came to the elk, they would get into one of the bands bunched up in the heavy snow and shoot as long as they had any cartridges left in their guns. Then they would pry the teeth out of their victims and leave the carcasses there for the coyotes to eat. Many wounded animals escaped, to die a lingering death. We wanted to stop that massacre.

It was a cold, rough job, with temperatures going down to twenty below zero. We operated with horses as much as we could and to some extent with snowshoes. Since the poachers kept watch on our comings and goings from the ranger station, we finally had to resort to leaving two or three men in the woods all night in a bivouac in the snow and subzero weather. That was the only way we could surprise and catch them. Those of us working on the Gallatin River had no exciting, wild experiences, though we finally arrested a leader of the gang hunting in that area. But there was no gun play—there wasn't any chance for it. We surprised him and had him dead to rights within the protected area where all killing of elk was forbidden. He didn't have a chance to resist arrest. In due time, he went to prison for his tooth hunting.

Over in the Yellowstone drainage, the law enforcement officers exchanged shots with a group of tooth hunters, but at long range, and no one was injured. However, our protective measures, combined with the activities of the Elks Lodge prohibiting the use of elk teeth as emblems, pretty well did away with the business of tooth hunting. That was the end of that. And it was also, as it happened, the end of my brief activity in game law enforcement, except for incidental supervision of enforcement activities on the Deerlodge National Forest when I was supervisor.

In connection with game protection, it is fair to bear in mind that in 1906 and 1907 when the Forest Service took jurisdiction over the western national forests, game laws were not very restrictive and were more honored in the breach than in the observance. In fact, the population as a whole did not give a great deal of thought to game laws. Yet, if an outside sportsman—so-called—came in and was destructive of game animals, everybody, ranchers and other local citizens alike, was glad to see him apprehended. However, a homesteader living out in an isolated region, struggling to develop and patent a piece of land, ate game. He wouldn't kill does and leave their fawns to die of starvation, but he would kill bucks. His killings did little to reduce the population of deer. I am forced to say that most of us forest officers were quite careful not to run across him when he had game. If we had encountered him with illegally killed game, we would have had to arrest him. We had a lot of sympathy for him.

In fact, in the first two or three summers that I worked for the Forest Service, when we were in wilderness country, backpacking, and often running short of food, I'm afraid that we didn't consult the law books before we knocked over a few fool hens to have something to eat that night. On one occasion when we were very short of food and had no prospect of getting more, we killed a young buck deer. Conditions justified it, and we were not killing needlessly.

The first winter after I was discharged from the army in World War I, I was stationed up in the Bridger Mountains. Fresh out of the army, I had very little money, but I had to put in a winter's supply of food because the road closed with snow. My assignment was to stay there all winter, snowshoeing between two timber sales on Brackett and Bridger Creeks, marking and scaling timber, and generally supervising the sales. I had a game license, of course, and it was hunting season when I went up there, fortunately, so I killed an elk to augment my meager supplies and put it in the deep freeze. My woodshed served admirably for that; it was stout enough to protect the elk from hungry coyotes, and the refrigerating

Catching a fool hen for dinner with a stick.

mechanism was the climate in the Bridgers up in Ross Peak Pass. That's cold country in winter, so I had no trouble keeping the elk at an appropriate temperature. I never had to call a repair man for my freezer. However, I must admit that at the end of three months a steady diet of elk with baking powder biscuits—dough gods, we called them—quite satisfied my appetite for elk. It's an excellent meat, but I had my share of it.

A friend of mine remembers that I took him to lunch at the Silver Bow Club in Butte one day years ago, and on the menu was a heading, simply, "Meat." We ordered it and ate it. It bore a striking resemblance to elk meat, although I doubt if our testimony would have held up in court. We would have had to qualify as experts on elk meat. Perhaps, though, I could have qualified as an expert after my three months of that diet.

One spring, a party composed of Forest Service, National Park Service, and State Game Department officials made a trip together in the country north of Yellowstone National Park. The purpose was to study the condition of the range and the condition of the elk comprising the northern Yellowstone herd, a continuing and difficult problem. Presently, the party came upon a black bear, recently emerged from hibernation. One of the men wanted to take a picture of the bear, so the party dismounted, left the horses with one member, and formed a circle around the bear. Becoming alarmed, the bear started to run, and, surrounded as it was, it inevitably ran toward one of the party, who happened to be the forest supervisor. To escape the bear's "charge," the supervisor instinctively headed for a small lodgepole pine tree with the intention of climbing it and escaping the bear. He was handicapped by heavy clothing and chaps but nevertheless struggled manfully to escape while the bear continued to gain on him.

Just before he reached the tree, the bear caught up with him—and swiftly passed him and climbed the tree. A newspaper reporter who had joined the party to get a wildlife story made the most of this. He had his wildlife story. It was a long time before the supervisor lived down the story of his race with the bear.

Even forty years ago*, the matter of suitable and sufficient forage for both wild and domestic animals presented a problem. In winter, cattlemen became irate, naturally, when hungry elk locked out of their own grazing lands by deep, heavily crusted snow started pawing and tearing down haystacks on the home ranch grounds of the cattle. The sportsmen complained that the natural winter feeding ground of the elk had been

---

* In the 1930s

eaten off by cattle in summer, leaving the elk no place to go. In summer, the sportsmen also protested our granting permits for cattle and sheep grazing on certain lands, terming it "the encroachment of cattle and sheep on elk grazing territory." Enthusiastic groups of sportsmen would get permission to trap and transplant elk from the too large Yellowstone herd to areas where there were no elk, but they seldom bothered to determine that there was winter forage available in the new location—and when the herd bred up to a large number, there was another controversy. It seems to be a continuing problem.

A beaver. —Photo courtesy National Park Service

CASTOR THE CONSERVATIONIST

Warm, sunny periods often interrupt the winters of our Montana mountains in late February and early March, bringing false hopes of spring. It was on just such a day in early March many years ago that I left my cabin in Ross Peak Pass in the Bridger Mountains and snowshoed down the little stream to where there was a beaver pond. At that elevation there was still deep snow, but the sun had softened it, and it was a soggy mess under my webs. In most places the brook was still buried deep under snow and ice, but at the lower end of the pond where the beavers had blocked the stream to form the pond, running water could be seen. The warmer water had cut the edge of the ice and opened a small space between the ice and the shore of the pond at one end of the dam. A trickle of water spilled over the dam and dived back under the snow.

Suddenly a dark head thrust itself out from under the ice and for a moment was still; then a beaver climbed a little farther out where the water spilled over the top of the dam. It carefully inspected the dam, and then sank out of sight. In a few minutes, it emerged again, this time with a water-logged stick, which it carefully adjusted in the dam. So far as I could see, this had no effect on the flow of the water, but apparently it satisfied the beaver, and certainly it was a more experienced dam builder than I. It climbed out on the dam and surveyed the world, then clumsily worked its way a few feet through the snow to where a small aspen grew on the shore at the end of the dam. This time there was no pause. Eagerly it began to chisel large chips from the trunk in an obvious intent to fell the tree. It was not hunger that impelled it, for it never paused to sample the tender inner bark. Perhaps, like me, it just had spring fever. More likely it was obeying the compulsive urge of its kind to gnaw something, and if there was a purpose, it may have been to gather material for the dam. However, the soft snow broke under one of my snowshoes, and as I caught my balance, the slight movement startled the beaver, and it plunged back through the snow and submerged, not even giving the usual warning slap of its tail due to lack of space.

This was the beginning of my acquaintance with this colony of beavers, for I had come to the cabin after the snow had locked them in. Later, when spring really arrived and the snow and ice had vanished, I met the rest of the family. They are generally nocturnal animals, but the high water threatened their dam and they were working a day shift. Mornings and evenings I could see them busily cutting willows and poplars and rushing them to the dam to strengthen impending breaks and to repair leaks. Since my work took me away during the day, I seldom had an opportunity to watch them in midday. Nevertheless, they shortly became accustomed to my presence and showed little alarm so long as I moved quietly. A sudden quick movement on my part and they would dive, slapping their tails as they went to warn the others. One warm evening, I strolled down by the pond and came on a large beaver out on the bank eating the tender spring vegetation. I stopped and it stared at me with myopic intensity, apparently interested but not much alarmed. Then it ate some leaves of clover before turning and making its way in a corpulent waddle back to the water. It didn't even bother to slap its tail when it finally dived.

As a forester in the Northern Rockies, I had seen many beavers, but this was my first opportunity to watch them closely over a considerable period of time. Let it be understood that I am not a zoologist and my observations and conclusions claim no scientific accuracy. However the pond, the dam, and the rodents themselves—known to the scientific community as *Castor canadensis*—afforded me a fascinating study. The stream on which the dam was built was a small one and the dam itself was not more than about one hundred feet long from one bank to the other. It was not a simple structure, a single dam. The main dam was nearly straight across the little valley. If it had been larger across a more rapid stream, they would have built it with a curve upstream, the better to resist the pressure. On the downstream side of the dam were a series of lesser and lower dams tied into the main dam and to each other, and each enclosed a little shallow pond of its own—in some cases they were only a few feet square. Thus the overflow from the pond did not escape in one spillway to dash down and perhaps undermine the dam. Rather it filtered through a maze of passages and little basins finally to be gathered in one channel. Willows and a growth of sedges and herbs grew on all the dams, binding them together and strengthening them with their intertwined roots. The instincts of these rodents had enabled them to construct a masterpiece of engineering.

The main pond was not a large one, probably less than an acre in area. Neither was the water deep—probably not over six or seven feet at its deepest point. Well out in the pond were two lodges, hemispherical domes of sticks and mud perhaps ten or twelve feet in diameter at water level. The entrances to these lodges were below water. The pond was fringed by a dense growth of bushy willows for much of its perimeter and with them grew a few alders and red osier. A little farther back from the water on drier soil were scattered quaking aspens, many of them scarred by the teeth of the beavers. Wild roses and the yellow blossoms of shrubby cinquefoil lent color. Everywhere were the stubs from willows that had been cut, and the larger stumps of aspens and alders. As spring advanced, a dense growth of sedges, grasses, fragrant white bog orchid, cow parsnips, and other moisture loving plants filled every available space.

The pond was a haven for various kinds of animal life as well as for plants. Undoubtedly the shallow water was filled with a rich growth of plankton—microscopic animals and plants, the latter chiefly algae. Dragonflies and other insects hatched and hovered above the water. Attracted by the insects, red-wing blackbirds, robins, and other birds congregated here. There were tracks of weasels, but these elusive little animals never permitted themselves to be seen. One or two mule deer came down to sample the tender browse, and if this had been moose country, almost certainly they would have competed for the young willow, alder, and poplar shoots.

What I was watching here was our first reclamation service at work, the earliest conservationists busy at flood control and soil and water conservation. Unrestrained, these mountain streams would continually deepen and widen their channels, extend their erosion, and carry their loads of silt in the spring floods to the rivers, there to be carried to the sea or to block channels and add to the destructiveness of floods in the lower reaches of the watersheds. However, before the coming of white men every mountain stream had its colonies of beavers. All of these were obstructing the waters, providing settling basins to catch the loads of silt, and cultivating dense thickets of brush to halt the erosion.

Then came the Hudson's Bay Company, and the mountain men, the Northwest Fur Company, and a host of others, all intent on exterminating the beavers to supply the nobility of Europe with hats and furs. The European species had been almost wiped out long ago. Fortunately, they did not succeed in exterminating our species, but they did fairly well eliminate the larger colonies—the dam builders—and the animals that remained were mostly those living in burrows in the banks of lakes and the larger streams. Unmaintained, many existing dams washed out during spring floods. Not all was lost. Over centuries and millennia, ponds had filled with alluvium, forming level, fertile meadows in the mountain valleys. Covered by bushes, trees, and thick turf, these meadows resisted the floods and even though deep channels might be dug through them, they remained essentially unharmed. Today, many of them are the fields of small farms, providing fertile soil relatively free from rocks and gravel. Occasionally the farmers developing the fields have plowed into half rotted sticks, the remains of some old beaver dam.

The fur trade finally collapsed, partly from the depletion of the supply of available fur and partly from changing styles. Beaver hats were no longer the thing. Some scattered sporadic trapping continued, but nevertheless the population of beavers grew steadily. Today with the passage and enforcement of game laws, we again have a large, perhaps adequate, number of beavers. In places they have become pests. They build dams that flood roads. They block the heads of irrigation canals. Dynamite a dam that is causing road flooding and within a few days, perhaps even the next morning, the dam is rebuilt and water again flooding the road. Clean out the headgate of the irrigation ditch and it certainly will be blocked again the next morning. I once operated a logging flume for transport of mining timbers near Butte. At intervals, "feeders"—small lateral flumes—led to the main flume to add water to replace that which had leaked out. Every morning it was necessary to dispatch men to clear out the intakes. About the only permanent solution when beavers interfere with man's activities is to trap them out, and it is from such trapping that much of our commercial beaver fur comes now. However, the fur is no longer very valuable; beaver is no longer in style.

In 1936, the federal Soil Conservation Service issued a pamphlet entitled "Little Waters."* In it they described and advocated the building of a large number of small dams and reservoirs on the "little waters," that is, on the small tributary streams at the headwaters of our major drainages. They pointed out the benefits that would accrue from controlling runoff and thus floods at the sources, their utility in storing water for irrigation and other purposes, and in raising the water table. The Soil Conservation Service must have been watching beavers. That is just what the beavers have been doing for centuries, though they were little concerned with the downstream benefits of their work.

---

*Little Waters: A Study of Headwaters Streams and Other Little Waters and Their Use and Relations to the Land.

Like men, they have a compulsive urge to modify any territory they occupy. But, unlike men, all they do seems to be to the benefit of other forms of life, animal and vegetable. I know no animal that they harm, and though they fell many trees, the net result of their work is to add to the density of the vegetation. Anyone who has fought his way through a

Divide Creek flume, on the Deerlodge National Forest in 1916, was used for transporting (fluming) mining timbers from the woods to the railroad. The logs were floated in long troughs that carried running water.

beaver swamp can testify to this. French peasants often "pollard" trees along the roadside—that is, they cut out the tops and harvest the resulting clusters of shoots for small sticks. The beaver does much the same thing, for the species of which it is fond sprouts freely and its cuttings are usually replaced by many more sprouts.

Obviously, a small pond, such as the one I have described, cannot support an indefinite number of animals. Each year some of the young must migrate and hunt a new home. During this dispersion they must be exposed to critical danger. Away from deep water or from its lodges or burrows, the beaver is defenseless. It can neither run nor fight. I have read statements that the beaver is seldom found far from water, which is true, but an associate of mine told of having encountered a beaver a mile from the nearest stream, crossing one drainage to another. I myself have found evidence of their work as much as a quarter mile from water. Such a beaver must be a sitting duck for any coyote, bobcat, or wolf.

Like most rodents beavers increase fairly rapidly. There must be some way of accounting for the fact that in nature the population soon becomes stabilized. There is little evidence that shortage of food in winter is a major control factor. It appears more probable that predation during the dispersion period is important in holding down the population.

Many people assume that all beaver live in lodges built in ponds, but this is incorrect. Many occupy burrows excavated in the banks of rivers and lakes. Their cuttings are in evidence remote from any pond where a lodge could be built. One summer I was fishing along the bank of the Kathleen River in Yukon Territory when a beaver came swimming upstream against the rapid current. It passed within ten feet of me and about fifty feet upstream dived and disappeared. I watched but never saw it emerge. Probably the underwater entrance to its burrow was close to where it submerged. Certainly there was no place in the vicinity where a conventional lodge could have been built. I wonder how they manage to excavate burrows and survive this far north. This was country of permafrost, though close to the southern boundary of permanently frozen ground. Quite likely in this comparatively sheltered spot along the river, the ground was not permanently frozen, but it must have afforded a cold shelter in winter in an area where the thermometer never rises above zero for months at a time.

Beavers have long been an inspiration for myths and legends. They are credited with remarkable wisdom and industry. Everyone has read tales of the man, fleeing from his enemies, who has dived into a beaver pond and entered a beaver lodge. Usually these stories have the beavers playing the part of hospitable hosts and welcoming the intruder. They make interesting fantasy until one has seen the entrance to a lodge where a pond has been drained. Then he is disillusioned by the size of the passage. It would be difficult for a small boy to enter, and if he did succeed in entering, you may be assured he would have the place to himself.

I doubt, too, that the beaver would rate very high, even among rodents, in intelligence. Certainly the common rat must have had to develop greater cunning in order to survive. Some of the dams are, indeed, remarkable achievements. Nevertheless, they are the result of instinct and not of planning, and there is a lot of trial and error involved. Many are started that are never finished, perhaps because the water was too swift, or for lack of suitable material close by. Moreover, beaver dams are not infallible. In seasons of unusually high water many break, but these are likely to be rebuilt. Any little leakage or break in a dam that lowers the water will bring a rush of activity and if at all possible repair will quickly be effected.

As for industry, the beaver has no alternative to frequent gnawing. His incisors continue to grow, and if there were such a thing as a lazy beaver, its teeth would shortly grow to such a length that it would be crippled. Occasionally they even fell pines and other conifers, which are of no use to them except to furnish an outlet for their obsession to gnaw. Their constant felling of trees and brush is merely the response to irresistible compulsion. They work for the same reason they breathe; they can't stop. Remember, too, that we see beavers only when they are out on the job. We have no way of knowing how much time they may idle away in their lodges. True, they do not lie on the bank in the sun and sleep. That would be an unhealthy form of recreation in a wilderness populated with wolves, bobcats, and mountain lions, which aren't lazy either.

Even though the beavers may not be overly intelligent or industrious after all, their ponds are things of beauty. A pond is a little ecosystem in itself, an enclave of water-loving plant and animal life tucked away in an austere mountain canyon. The focus of the ecological association is of course the pond itself. It is a haven for young trout, for here in the sedges, rushes, and pond lilies, they can find shelter from the large dolly vardens and other big fish. There is a rich profusion of algae, protozoans, and insect larvae to provide ample food. Families of waterfowl hatch their young and harvest their quota of aquatic life. Bog orchids, buttercups, marsh marigolds, and the coarse leaves and white umbels of the cow parsnips border the pond, competing with the sedges and redtop for every open spot in the thickets of willows and alders. Frequently a moose can be seen, knee deep in water, feeding on yellow pond lilies. There is predation but there is also mutual interdependence, a sort of symbiosis. Here one can come about as close as possible anywhere to observing that mythical ecological phenomenon, the biological balance. Of course there is never a perfect balance; it is continually being upset. But if some species begins to achieve dominance, it tends to attract some other to feed on it. The point of balance is always swinging back and forth, but the zero mark still remains and the community is seldom far from it.

I have already stated that I am merely a casual observer of beavers. One may well ask why I do not consult some of the reports of studies by competent scientists to find the answers to my questions and to correct my probably incorrect assumptions. The information undoubtedly is readily available. The answer is that I simply do not choose to do so. Mark Twain in his *Life on the Mississippi* explains why. After telling of his training as a river pilot, he wrote: "Now when I had mastered the language of the water, and had come to know every trifling feature that bordered the great river as familiarly as I knew the letters of the alphabet, I had made a valuable acquisition. But I had lost something, too. I had lost something which could never be restored to me while I lived. All the grace, the beauty, the poetry, had gone out of the majestic river! . . . The romance and beauty were all gone."

Foresters, too, lose something of the romance and beauty of the forest. All too often, I have topped a ridge to look down into a mountain basin and seen not the grandeur and beauty but a difficult problem in the control of some possible future fire. The veteran pine that once I would have regarded with awe and reverence has a bracket fungus, which indicates that it is decaying and is doomed. That young grove of trees is infested with mistletoe and should be thinned. But it is not essential to my work as a forester that I increase my understanding of the complex relationships of the beaver swamp. I have no need to pry into the beaver's domestic arrangements or to intrude on its privacy. I wish I could go back to my youth and could believe in the beaver's wisdom and superlative engineering skill. I would be happy to regard its lodge as a sanctuary for some hero, fleeing from his enemies. However, those illusions are gone.

But there still is much of mystery to add to the attraction of the scene. I shall cherish that mystery. I have camped by many of these ponds. In the still of the evening the mirror surface is broken by a V-shaped ripple. Here comes the promoter and builder of the community, a beaver busy on some errand, I know not what. A deer ventures down to drink and to crop some of the young twigs. Far up the pond a loon breaks the silence with its wild, lunatic, quavering cry. This is my Walden Pond. Let it remain so.

WOMEN OF COURAGE

Too much cannot be said for the courageous wives of our early-day foresters, especially the wives of our rangers. Our stations for the first few years of the Forest Service were pretty crude affairs. As I remember, even as late as 1921–1922, the limit established by Congress for expenditure on a ranger station dwelling was $650. We could, however, use the labor of our Forest Service officers in the construction. But $650 was just barely enough to put up a three- or four-room log cabin by using every economy. We didn't have appropriations in the earliest years to put up additional bunkhouses for laborers. There were no stopping places for inspectors out of the supervisors' offices or from the regional office. They expected to have meals and lodging at the ranger's home. At times the ranger's wife found herself running a boarding and lodging house.

Moreover, the hastily selected sites of some of the ranger stations were unfortunate. Late one fall, in a period when work was slack, the foundation was laid for a station on the Yellowstone River on the Absaroka National Forest at the foot of Dome Mountain. The building was finished before winter came, enabling the ranger with his wife and children to move in. When spring came with warm weather, they discovered that the station had been built at the rocky foot of a mountain harboring a rattlesnake den, where the snakes gathered for the winter in crevices in the rock. Of course, the pests were reduced after a while when the snakes had scattered, but when they came out of winter quarters in the spring, the entire area, including the station yard, was literally lousy with rattlesnakes. The small children didn't dare go outside of the house and even a trip for a pail of water after dark was an adventure. Many a ranger's wife lived with these primitive or unsatisfactory accommodations.

Remember, the ranger was not an uneducated laborer, and his wife was not an illiterate frontier woman, accustomed to nothing better; she was usually an educated woman with refined tastes. Yet she lived in these crude cabins, even carrying her water from a spring or stream. During the fire season, when her husband was out battling forest fires, in addition to her regular duties, she took on many of the ranger's responsibilities, spending hours a day on the telephone handling forest business.

There is one famous case of a temporary summer employee going insane at a ranger station when the ranger's wife was alone there. The young man, who had possibly sustained brain damage while playing football, was housed in a toolshed back of the dwelling where Mrs. Ralph Fields, acting on telephone instructions from Ralph, was waiting behind locked doors. The boy cleaned off a tool chest, and Hulda watched him lining up dynamite caps on top of the chest and making designs of them—"enough to blow everything up," Hulda said. Then the boy bludgeoned himself on the head and finally stabbed himself in the stomach with a butcher knife and drank some Lysol he had found in a first aid chest.

Hulda phoned Ralph, who sprinted the four miles down from a lookout station, arriving at the ranger station at the same time as the doctor and the sheriff, also called by Hulda. She was beating raw eggs to get into the sick boy to counteract the Lysol when the men arrived. He died in a hospital. Hulda, a twenty-one-year-old girl, was then expecting her

The Carney Ranger Station in 1911 in the Madison National Forest.
—Photo courtesy US Forest Service

first child. Rarely was there anything this tragic, but this exemplifies the stout courage of the women who, in many respects, were the inspiration of our young men. Incidentally, this was Hulda's first experience in the deep woods, living several miles from the nearest neighbor. She was a sorority girl and had just recently earned her degree in journalism from the University of Montana.

The Forest Service used to send a supervisor each year from some western forest to tour a lot of the eastern schools in winter. One year I drew that assignment and a rugged one it was. For two and one-half months I traveled around to a dozen of the forestry schools, including the universities of Pennsylvania State, Yale, New Hampshire, Connecticut Agricultural, Maine, Syracuse, Cornell, Iowa State, Purdue, Michigan, Michigan State, and Minnesota. In addition to holding consultations with the faculty and giving organized talks to the students at the request of the schools, I interviewed students and, incidentally, gave talks when requested to such luncheon clubs as Rotary and Kiwanis. The university sessions gave students a chance to talk with a Forest Service man and to ask questions about the possibilities of employment and the conditions they would encounter.

Fleecer Ranger Station on the Deerlodge National Forest in 1923. —Photo courtesy US Forest Service

At the University of Maine, the forestry school chiefly supplied forest-ers for the spruce paper pulp industry in New England. Here, a young man, somewhat fascinated with the idea of going out west in the Forest Service, asked me a lot of questions and finally got down to the main one. He said he was engaged to be married and he wondered if a man could take a wife out into those conditions. I thought that over for a moment, then said, "Well, I did." He continued, "Yes, but could one take the kind of a girl a man of my background would want to marry?" When I reported this to my wife in great amusement, she came back with, "Well, of course, the reflection is on *your* background." At the time, I managed to keep my face straight, but I think the remark is worth comment.

The question he asked was one that I would be incapable of answer-ing unless I knew the girl, and probably not then. I don't believe it would make any difference whether she came from Maine or Montana. It would depend on the type of girl. Conditions at some of the places where rangers might be stationed would be just about as difficult for a girl from Missoula or Helena as they would be for one from New York City. It would just depend on her courage.

I remember one young man who married a sorority girl from an eastern university. He drew an assignment for the winter on game studies at the Moose Creek Ranger Station in the heart of the wilderness area of north-ern Idaho. This girl went with him. They were taken in in the fall on horses with their entire winter supply of food and were completely isolated all winter. The only possible way to get in or out would be about a two-day trip on snowshoes. That was a rugged, severe enough test, yet that young woman reported she enjoyed it immensely. She fitted into the conditions and made the best of things.

And then, I have known girls raised in both western and eastern com-munities who flatly refused to live out in some of those isolated places. Being a tenderfoot or a quitter isn't a matter of geography; it's a matter of adaptability and temperament. Of course, anyone can be a tenderfoot in the sense of being ignorant of local conditions, but that doesn't last long if he or she has the right attitude. That goes for both men and women. By and large, women proved to be at least as adaptable and courageous as the men, if not more so.

One season I had to take charge of a crew fighting a fire on the North Fork of the Flathead River just across the river from Glacier National Park. Arriving at the base camp, I found things in bad shape—the fire was crowning, crossing firelines and roads, and burning in Glacier National Park as well as on the national forest. I went to the phone to report to the supervisor (it was a party line and naturally before ringing I listened to see if it was in use) and heard a woman's voice. Something in her voice made me continue to listen. She was a wife of a park ranger, just a few miles away, describing the inferno that was roaring up the canyon.

The supervisor, Jimmy Ryan, asked, "How are you and your baby going to get out?"

"Well, Mr. Ryan," she replied, "it doesn't look like we're going to get out."

At that point I broke into the talk, identified myself, and told Ryan that I'd get in to her some way; the fire was spotty between us and the park station. Heading for a car, I picked up a National Park Service man who had become separated from his outfit. He said he was familiar with the country (I wasn't) and that it was his job to go. We commandeered a car—I never found out whose—and headed up the crude road, edging through the burning trees. Not far from a bridge across the river, the road was blocked by a fallen tree, and the park man (I never learned his name but he was all man) said, "Here's where I take off." He headed right into the flames. I backed the car down the road and returned to camp.

Then the fire did one of those strange things that fires often do. It divided for no apparent reason—I suppose some freak air current built up—and never did hit the park ranger station. The man got to the park station and made preparations to get the young woman and her baby to the nearby river, but kind Providence saved them from that ordeal. Park rangers had brave wives, too.

Experiences of that sort were not too unusual, as many a ranger's wife discovered.

One of the top priority jobs in the early years of the Forest Service was setting up a fire lookout board on a high mountain peak to assist in determining the exact location of lightning strikes and other incipient fires. There was no shelter—the board was out in the open. However, with a water-proofed map on the top, complete with azimuth circle and an alidade for sighting, this was an effective device until lightning struck the peak. Then they had to get another lookout. Usually, though, the seasonal employee lived in a tent or small cabin at the base of the mountain, climbing up each day to chart the fires. Cabins or towers fully protected from lightning strikes evolved only after many years of harrowing experiences. With Ranger Joe Callahan, I posted such a lookout board on top of 11,200-foot Long Mountain in the Gallatin National Forest in the spring of 1920.

BEHIND THE FIRE LINES

To recruit the thousands of men needed to fight the huge fires that confronted us at times during these years was a job of considerable magnitude. Our career Forest Service men, of course, constituted the overhead, directing the battles, often with a zealousness and driving dedication that eroded their own physical well being. But the ranks had to be built up from local laborers and seasonal workers—strong men who could handle shovels, picks, axes, saws, and hodags.*

The old transient laborer was a man of virtually no family connections. He was the fellow who worked in the Imperial Valley in the dead of winter, migrated to the hop fields of Oregon, followed the threshing crews in the Dakotas and Montana, and came out when we had a fire season to hire on as a firefighter. His kind has almost vanished. He represented a class that was socially undesirable. It was a big criticism of the state of our society that we had those drifting men who could not establish homes and families, and who had frequently lost touch with parents and relatives. One could tell of them for a long time. They were hired in this region predominantly along Trent Avenue in Spokane, Washington; in so-called jungles in Missoula, Montana; and down on Arizona and Mercury Streets in Butte, Montana, an area of cheap flophouses and dives.

It was from that area in Butte where, during my years as supervisor there, I had to recruit these men to ship to the big fires on the forests farther west. This was an area that the chief of police, the famous old Jere Murphy, refused to police. He said that nobody had any business down there if he was a decent man, and that he wouldn't waste his cops in that part of town. He advised me, "When you go down there to hire

---

*A hodag is a mattock with one axe-bladed edge for cutting roots and one edge set like a hoe for digging a trench around a fire.

firefighters, carry a gun, and if they look at you crosswise, let them have it." So I did carry a gun, although only twice did I have to show it, and I never had to use it. Even that was probably a needless precaution.

Oddly enough, those itinerant laborers, bitter as they were against society generally, often were exceedingly loyal. We fed them well. We had excellent food in our camps after the first few years. The men knew if they were injured we would care for them, a change also after the first few years. That care went beyond merely putting them into the hospital and paying the bills. The Forest Service had a strange solidarity, and these firefighters who were hospitalized would receive gifts of cigarettes and magazines from staff members who would take time to go to see them and to check on their care. In short, so long as they worked for us, they were part of the Forest Service and were treated accordingly.

We found all sorts among those crews. When we hired them, we asked them to designate their next of kin whom we could notify in case of an accident, and a surprising number answered, "I have no one." They had no one who cared whether they lived or died. Two of them gave me the name of the mayor of Seattle—that city's first woman mayor. They said perhaps she'd like to know if they were killed, because she had kept them in jail out there on vagrancy charges for three months. They took the matter rather lightly. But they were skilled with fire tools and loyal under good leadership.

One day when I was rounding out a crew in Butte to send over west, a man came into my office whom I knew, a department manager at Symons Department Store. With him was a strapping young fellow, immaculately dressed, who was introduced as a nephew from New York City. It seemed the young man was out west with his bride and was look-ing for adventure, so he wanted to hire out as a firefighter. I explained to him that firefighting was just hard, smoky, back-breaking work. I told him about the living conditions—sleeping on the ground and eating at irregular intervals, though I had to admit that the food was first rate, at least up to the time the cooks got it. I pointed out that he didn't have the right clothes—loggers boots, wool shirt, Levi pants, wool socks, and leather gloves. The uncle said he would outfit the young man, so I hired him against my better judgment.

He was on time at the train when it came time to ship out and he had the proper work clothes. But I expected him to fade in less than three days. Several weeks later the rains came and the fire season at last folded. Then the young fellow showed up, tanned, work-hardened, his clothes scorched in spots—and the boss of that crew of itinerant laborers. No, you can't always tell about men.

Wildfire in 1929. —Photo by K. D. Swan, US Forest Service

Another time I shipped out a crew and a day later two well-dressed women came to the office to inquire about two boys and to ask if I had hired them. I checked and found that I had.

"But," said one of the women, "couldn't you see that they were gentlemen?"

I admitted that I hadn't noticed it. She then asked if we had suitable accommodations for gentlemen, separate rooms, and if there was suitable food. I replied that I never had trouble eating the food, which didn't seem to reassure her. And, as far as separate rooms were concerned, there was a million acres, more or less, on that forest, so the boys could have all the room they wanted. We were not surprised when the two "gentlemen" came home in a couple of days.

The best firefighters I ever had were miners; the choicest crews were the Finnish men. They are, as a group, men of remarkable physique and proud of their physical prowess. However, they require special handling. I didn't attempt to ship my Finnish friends to other forests, after several efforts. They simply wouldn't work under the lumberjack type of leadership. In the first place, my Finnish crews always picked their own foremen. Usually, if I called up a boardinghouse in the Finnish part of Butte and asked them to send a twenty-five-man crew for some fire, I would recognize the foreman as one who had served in that capacity before. But if I didn't recognize a foreman in the group, I'd ask them who their foreman was. They always followed their own leader, so I found it expedient to appoint him in the first place. I valued them because they were wonderful firefighters and high class men. Of course, most of them had learned their woodsmanship in Finland. As there were only a limited number of them available, I came to use them only for fires on the Deerlodge National Forest.

I had quite a fire season on the Deerlodge in 1929, which kept me going from fire to fire to check on operations. On the Gray Eagle Fire, I noticed a fellow who looked like anything but a conventional firefighter. He looked out of place, so I asked where he was from. He replied, "Chicago." When I asked what on earth he was doing out here on a fire crew, he explained his predicament.

The fellow said, "Well I was in the rackets back there, but I got in bad. The boys were going to take me for a ride, so I had to leave town. I thought I would go west and find myself a good live town and start up a racket on my own. I heard that Butte was a good, live place, so I came there. And, you know I wasn't in that town two hours before they had me up before that Irish chief of police, Jere Murphy, and Jere said to me, 'Boy, we have a nice little prison down at Deer Lodge, but that isn't for the likes

Aftermath of the Gray Eagle Fire on August 23, 1929.

of you. That is for the home boys. You better get out of town or you'll be bumped off.' Well, the quickest way I could get out of town was to hire on with you people."

About this same period, a young ranger who had just finished work on an insect infestation job in the Big Hole came into Butte on his way west. Having heard about the town's nationwide reputation as a wild and riotous place, he was eager to see it. After dinner at his hotel, the Finlen Hotel, he headed down toward the Cabbage Patch, a rough district. Having passed out of the respectable area, he went only about half a block until, passing an alley, he was held up and every penny he had (some eighty dollars) was taken off him. That kept him from catching his train next morning to go back to his headquarters. He had to wait until the Forest Service office opened to get me to identify him so he could get some money. Likewise, he felt that he should properly report the holdup to the police. I told him I didn't think they would be interested, but nevertheless I took him down to Jere Murphy and he told his story.

Jere looked at him and finally said, "Well, boy, and what were you doing down in that part of town?"

The boy had no answer. He hadn't followed the code: if you want to stay out of trouble, stay in the decent part of town. Butte presented many difficult problems at that time, but it was generally agreed that its nationally famous chief of police maintained about as good order and protection for its citizens as was possible under those conditions. Years before, when I had arrived in Butte after an extensive period out in the woods where I had no news of what was going on outside, I got off the train dressed in rough woods clothes and with a Winchester rifle and was promptly grabbed by a soldier and held for questioning. What a surprise! Butte was under martial law because of labor disputes. However, I was soon cleared and permitted to go about my business (minus my rifle, which I retrieved when I left town.)

Every big fire season when we were hiring large numbers of firefighters, someone would try to put out fraudulent checks. We had one case in which a lot of blank government check forms were stolen from Forest Service headquarters at St. Maries, Idaho. According to the FBI, which investigated the crime, there was a skilled penman involved who made out the checks and forged the fiscal agent's signature.

During the 1929 fire season, I had a phone call one day from a dive keeper in Butte, inquiring about a check he had accepted that bore what was supposed to be my signature. It was presented by a supposed firefighter. Of course, I did not write government checks; that was our fiscal agent's job. The dive keeper brought the check down to my office. It was for $125 with my signature and title. Of course, it wasn't a government check, and the signature bore no resemblance to mine. I had been very busy, and my wife had been out of town. This phone call alerted us to check our bank statement that had just arrived in the mail. We were shocked to find that we had no account left. A whole flock of forged checks had carelessly been accepted by the bank, putting our account officially in overdraft. Somewhat red-faced, the bank quickly corrected the situation.

The dive keeper wanted to report this to the police and asked me to go down with him to make the report. Jere looked the check over quite thoroughly. Then he called in an assistant.

"When did so-and-so get out of Deer Lodge?" he asked. "I thought he was safe down there."

"No, he was sprung a month ago."

"Well, he's at it again, passing checks," said Jere. He had immediately recognized the work as being that of a repeated forger.

"Did you lose any money, John?" he asked.

"No, the bank's responsible for that; it wasn't my signature," I replied.

"Well, I expect they could stand it. Where were they cashed?"

"Oh, they were all passed in saloons and dives around here," I answered.

He looked at the check again, commenting, "Yes, and those endorsements will all be good. They're in that business; they can afford to lose it. Well, if we happen to see him, we'll pick him up, but it doesn't make any difference whether he's in or out." His attitude was that as long as a crook preyed only on crooks, it wasn't a matter for decent people to worry about.

We have heard many stories of the adventure and heroism in the fires in 1910 in northern Idaho and western Montana when almost eighty Forest Service employees were killed. Several thousand others were injured. The heroism of Ranger Ed Pulaski became almost legendary. He went into the face of a raging crown fire on the Coeur d'Alene divide to rescue a crew of firefighters. It was too late to guide them out, but he herded the panic-stricken men into a mine tunnel at the point of a gun and hastily erected a sort of bulkhead to let the fire go over them. One or two men died of suffocation. Ed, himself, was in the hospital for a long time. Injured members of the crew were also hospitalized.

What is seldom told is that the government's only official concern at that time was to make certain that a man's pay was stopped on the day on which he became incapacitated. The government did not pay one penny to bring them out or to hospitalize them. Those who did not have resources of their own had to go on charity. As many of these laborers were in that category, the nearby communities were burdened with the care of these helpless, injured men. Ed Pulaski, the hero, had to pay his own doctor and hospital bills.

Some twenty itinerant laborers were buried in a trench in the St. Joe National Forest, two days by pack train from the nearest road, wrapped in Forest Service sougans. (A sougan was a cheap comforter popular with sheepherders and also usually a part of the lumberjack's portable bedroll.)

It should be said that, with generosity, that next winter of 1910–1911 Congress made an appropriation to reimburse some of these hospital expenses. It also provided for bringing out the bodies buried up there in the woods and for purchase of two cemeteries for the Forest Service, one at Newport, Washington, the other at St. Maries, Idaho. The firefighters were decently buried there. In subsequent years, we added others to those little cemeteries, and I suppose they are still maintained by the Forest Service. Moreover, legislation shortly provided for compensation for injury in line of duty, care of those injured, and burial of those killed while in the Forest Service of the country.

Soldiers from Fort Missoula working the Hay Creek Fire in 1926.

SOLDIERS IN A PEACETIME WAR

I had a battalion of regular army troops on a fire west of Glacier National Park on the North Fork of the Flathead River in 1926. They were sent there because the fire threatened not only the national forest but also Glacier National Park. This battalion consisted of the old professional soldiers staffed by West Point officers. They were unskilled with firefighting tools, but they worked hard and willingly and were a good crew. My main concern was that they were too literally obedient to orders. It's difficult to put in sufficient supervision to assure the safety of every man on a fire line unless the man himself uses a little discretion. He should have sense enough to run when the time comes to run. Soldiers were drilled to hold a position until ordered elsewhere; they expected immediate leadership. That is a valuable trait against human enemies in war, but a poor trait on a forest fire, especially when it starts crowning. However, the immediate leaders, the officers, did not recognize the indicators of danger. Nevertheless, it was an efficient crew.

The major in command happened to be an officer under whom I had served briefly as an orderly when I was an enlisted man during World War I. I had the feeling that he felt it was somewhat beneath his dignity to be under the direction of a former enlisted man. He conceived the idea that the way to fight fires was to establish his forces in echelon and build second lines to which he could drop back if driven out of the first line. This may be good military tactics, but it doesn't work too well against a forest fire. However, he was so insistent that I felt compelled to permit him to take part of his troops to put in a secondary line back of the main fire line. This, of course, was built up very close to the fire where we could back-fire to make an effective barrier to the build-up of the fire.

This was a difficult stretch of country with a very difficult fuel type, and I had little hope of holding the fire. But we had to do our best. One day, while the main group was still building line, the weather forecasts

Hay Creek Fire on the North Fork of the Flathead River on the Flathead National Forest in 1926. —Photos courtesy US Forest Service

and all accompanying conditions became so critical, in my judgment, that I pulled the troops off both my line and the major's secondary line. I got them out and into the clean burn just in time. The fire picked up and swept through everything and nothing on earth could have stopped it. Ten thousand men would have been just as helpless and, in fact, would merely have been killed if they had remained on the line. When the fire reached his secondary line, it never slowed at all. When a fire is crowning in heavy coniferous timber—racing through the tree tops—it will leap a fairly broad river and never even slow down—it's just a mass of flame, heat, wind, and roaring—an awesome and terrifying sight.

Even that didn't teach the major a great deal about firefighting. Having lost that line, we backed up and established camp in a forty-acre homestead clearing from where we attempted to establish another control line. However, weather conditions continued very unfavorable, and it was apparent that we weren't even going to get well started. So I ordered camp moved back up somewhat farther, to which the major objected strongly. He was positive that in that forty-acre clearing we would be entirely safe, but I assured him we would not.

I had to become very insistent. In fact, I reminded him that in the army one was expected to obey orders, that I was in command, and that I wanted to know if he were going to obey my orders. He angrily agreed to do so. We were near an old road where we had trucks, so we could move fast; but we got out none too soon. The fire nearly caught the last truck. At a predetermined location, we left the trucks. Then he issued the orders, and the whole battalion moved into the burned over area where there was nothing combustible left. The fire swept through on a run that went across the river into Glacier National Park.

I led the way into the burned area, needling our way through spots of still burning snags, over blackened, smoking terrain. As I looked back on the column of men to see how they were coming, I saw those disciplined soldiers, scared stiff, but marching at attention with perfect discipline. I hollered back, "Relax men, you're safe." Later I learned that the *Missoulian* had large, red banner headlines across the top of the front page that July day in 1926: "Taylor and 200 Men Feared Lost." That newspaper was brought to my bride to be as she was trying on her wedding gown.

That was a very destructive fire, burning through thousands of acres, but by that time we were in a safe spot. The next morning, I took the major to the site where he wanted to camp—the forty-acre clearing. The fire had swept across the clearing with a blast that knocked down trees two and three feet in diameter, on the far side. We would all have been killed instantly had we remained there. And let me do the major credit. After we

A pack string passes through land burned by the Hay Creek Fire in 1926.

returned to camp, he called the other officers together and in their presence apologized to me for his intransigence the day before and admitted
he had been wrong. This wasn't necessary, but it was a tribute to his moral
courage and decency.

This was the fire where, in deference to military practice, I departed
from our usual custom and set up an officers' mess. This merely consisted
of an area apart from the firefighters where the officers could eat. They
were given no other preference. I was with the officers. At our first dinner
after the troops arrived, the dessert happened to be large, luscious, high-
grade canned peaches. (We served excellent food in those fire camps.)

The major said, "Why Mr. Taylor, you didn't need to get food of this sort
for us officers; we are ready to eat just what the other men eat."

"That is just what your men are eating, Major," I replied.

"My God," he exclaimed, "Melba peaches for enlisted men—it'll ruin
them!"

About 1932 we had another bad fire season with fires going strong on
the Cabinet National Forest in western Montana. One of the big problems
when you have a lot of men out in the woods is supplies. Everything
they have back in those areas—food, beds, equipment, miscellaneous

supplies—has to be purchased, assembled, and sent in proper quantities and at the right time. The quantities handled come as a shock to anyone unaccustomed to supplying large numbers of men. That year we had a man, Dick Hammett, later known as the father of Smokey Bear, who had just come to Region One. His previous experience had not involved fires of such size; he had been supervisor of a forest in northern California, where he had never encountered one of these project fires. In deploying the professional overhead, the fire chief detailed him to dispatch supplies, a job for which he was well qualified, but the quantities of food required amazed him. One day he was receiving a long order for food and the following conversation ensued:

"And we want ten dozen cases of eggs."

"You mean ten dozen eggs, don't you?"

"No, ten dozen cases of eggs."

"A case has forty-eight dozen."

"We know it. We want ten dozen cases of eggs."

"My God, what a job for the hen!"

Actually, for many years the fighting of fires was a frustrating and discouraging battle. We saved a lot of country, we cut the fire off from advancing on this and on that front, we restricted the size of the burned areas, but we often didn't extinguish the last of the very large conflagrations until help came from rain or snow in the fall. Fire was a constant worry to all professional foresters from June 1 to the end of the fire season in fall. One was continually in reach of communications, tensely alerted to a possible fire detail fighting fire for the rest of the summer anywhere in this region or in some other region that was in trouble and calling for help. Inescapably, there was confusion at times, because communication and transportation facilities were very poor, often bordering on being haphazard.

Early one morning, veteran pilot Bob Johnson was flying a group of us to Grangeville, Idaho, to help on their fire staff there. Our route was over the Selway Wilderness. With a mixture of smoke and low clouds to complicate conditions, Bob had to follow down the canyon, threading above the Lochsa and Clearwater Rivers. Presently, visibility became obscured until we could see neither the river nor the mountains on either side. Then a momentary clearing revealed Pete King Ranger Station and gave us our location. But again the cloud cover closed as we flew out over the Camas Prairie, and by the time Bob managed to drop through a break in the cloud cover, we were momentarily lost. Bob, however, spotted the branch railroad and followed it to a station where he again got his bearings, quickly bringing us to Grangeville.

This flight was over country where, some twelve years earlier, I had got myself into a really tight situation, in August of 1919. Frances Carroll and I had scouted a stretch of uncontrolled fire, and we then foolishly took the easy instead of the safe way to return to our station. The fire went into the crowns and started a bad run with us out in front of it. Fortunately, we were near a patch where the timber was thin from some smaller previous fire. We awaited the approaching fire until it reached the lower end of this old burn, then charged at full speed into the face of the fire to try to get through the intense heat of the edge of it as quickly as possible, and into burned country. Needless to say, we succeeded and were uninjured, though we spent a very uncomfortable half or three-quarters of an hour working our way through smoke, heat, and falling snags to the river.

Uncomfortable experiences of this sort were not uncommon in the western part of the region, where the worst fires and most serious hazards were met. Even so, experienced men seldom got into situations they couldn't deal with. Casualties were chiefly among temporary laborers and were frequently due to panic and disobedience of orders. I tell of these experiences not because they were extraordinary, but because they typify the life of a forest officer during one of those fire years.

I have been asked how the recent bad fire year of 1967 compared with that of 1910. I suspect it is impossible to make a fair comparison except to say that both were dry years. I might point out interesting contrasts of three conditions. An enormous factor of the 1910 season was the inexperience of the four-year-old Forest Service, plus lack of equipment and manpower. We hadn't had a chance to put in telephone lines or to build roads and trails. Travel was largely over old Indian trails, or trails put in by prospectors, and was extremely slow. It was impossible under the existing conditions to contain those fires, largely caused by lightning but some by man.

Though we had a lot of fires burning in mid-August, we had managed to restrain them sufficiently so that none of them were especially large. Then, around August 19 or 20, an unprecedented windstorm swept through the country. It picked up the fires and whipped them together. As they gained momentum, they roared across Idaho and western Montana for over one hundred miles, burning up several million acres of timber. Burning embers were carried aloft by tremendous drafts created by the fire, to fall in towns many miles away from the nearest fire. Darkness came in midafternoon as the sun was obscured by dense, acrid smoke. Ashes sifted down over nearby cities, and communities were burned out. There are endless stories of the excitement and narrow escapes of those two days. Our feeble little crews were utterly helpless against that holocaust. It became a matter of trying to escape alive.

If the 1910 fires, at the time of the big wind, had been as large as some of those in 1967, forests would have been wiped out wholesale. On the other hand, if the modern organization were to face an equally terrific wind with an equal number of fires burning, even present techniques, skill, and facilities would be overcome. However, now, with airplanes, smoke jumpers, radio communications, more roads and trails, use of fire retardants sprayed from planes, and with bulldozers and other line building equipment, fires are extinguished quickly. It is very unlikely, therefore, that such a wind would find many fires burning and, consequently, it would not cause a disaster. I believe the burning conditions in 1967 probably were worse than in most of the 1910 season, and the loss was serious, but it was trivial compared with that of 1910. I doubt if we shall ever see such a fire in this region again.

A private cabin on the Deerlodge National Forest in 1929.

MOVING THE MOONSHINERS

While the Forest Service had no responsibilities whatever for enforcement of the Prohibition Law—that was not our business—we were interested in seeing that the federal land under our jurisdiction was not illegally occupied or devoted to illegal purposes. That definitely was our responsibility. However, the forests with their clear streams of good water, plus the necessary concealment features, provided an ideal place for moonshine stills. Consequently, we had many of them.

My encounter with the moonshiners took place around Butte and Anaconda in the Deerlodge National Forest. We kept finding stills tucked back in gulches around these cities on national forest land. We protested. The moonshiners became increasingly defiant, flatly refusing to move off. We reminded them that there was lots of private land even within the forest boundaries—old patented mining claims and homesteads—where their illegal occupancy would be no business of ours. But when they continued after all this warning to locate on forest land, it called for some action on our part.

We passed out word that if they did not move, we would call in the prohibition officers to raid them and dispose of the stills. Word came back that if we did anything of the sort, we'd get shot. That was just a bluff, but we couldn't permit ourselves to be bluffed. We called in the prohibition officers.

However, there was a saying that prohibition officers often were in league with the moonshiners and bootleggers. And, in some cases, we found that the raiding and destruction of the stills was quite inconclusive, so to speak. The officers were quite ready to conduct the raids; they had to make a showing. But when they got to the still, they would merely destroy the worm and perhaps smash one little hole in the copper still, which could readily be repaired by soldering on a patch. The worm, of course, was just a little piece of copper tubing that could easily be replaced. Very likely, also, the officers would throw a little distillate on top of the mash barrels. It was simple enough to siphon off the mash from under the distillate—that caused very little trouble—and in two or three days the still would be in operation again.

Of course, if the raid was conducted while the still tender was present, he would be arrested and put in jail for from three to six months, during which time his employer paid him as usual. He would even see that he was well fed on top of the jail rations. These months in jail became an anticipated and welcome diversion for the tenders of the moonshine stills, a paid vacation in a select club. In jail, a sort of segregation developed with the moonshiners housed in their own section of the jail—a Shangri-la as it were. It was reported that their special food service and bar, paid for by their employers, ranked with the best in the city. Since the sheriff received an allowance for feeding these prisoners, which he could pocket, the arrangement met with his hearty approval. Obviously, in more ways than one, the raids weren't helping the situation any.

The problem grew worse in the fall, just when we had had a spell of rainy weather. This enabled us to take control ourselves, in our territory, of course. After the officers had gone through their routine, we would open up the barrels of distillate and let the contents dribble around in the still house. Then we would drop a match into it. The resulting blaze completely eliminated the still, grain, barrels, mash, sugar, and all. That still never would operate again.

We knocked over and destroyed eight stills, each representing a substantial investment of up to six or eight thousand dollars. I haven't the slightest idea of how many we actually moved off the forest, however. Our tactics caused all those who were operating on the forest to seek other locations. One fellow came into my office during this period of sudden violence, very agitated. He asked me to delay raiding him for just a few days until he could move his still. I told him we couldn't make any agreements or promises. I said, "We'll take care of you when we get to you, that's all." Actually, we hadn't known of his still until then. One day, when going by his little ranch, I noticed a number of barrels stacked around there, indicating that he had probably brought his still in.

After this period of decisive action, moonshiners were careful to erect their stills on private land, generally on patented mining claims, on which we would not trespass. Our brief encounter with moonshiners was quite successful except in one case.

East of Butte was a park with pansy gardens and a playground for children, a widely known attraction called Columbia Gardens. It was a philanthropy for the city of Butte, donated and maintained by former US

Senator William A. Clark, the copper magnate. It had its own water system quite apart from that of the city, something I knew but completely overlooked at the time. One of the stills we destroyed was just above the intake to the Columbia Gardens water system. At Columbia Gardens that September day, they were having their last childrens' day of the season. This was always a special occasion, with all the children getting free streetcar rides out there and back, plus free rides on all the entertainment devices and the privilege of picking bouquets of pansies from the gardens. Naturally, the place was filled with children. That was the time, by chance, that we chose to dump several hundred gallons of moonshine and possibly a thousand gallons of mash into that little stream.

The first they knew of our action down in the gardens was when a little boy said to his mother, "Mamma, this water tastes like soda pop." It was running pure highballs out of the faucets at Columbia Gardens. The newspapers loved the story. I suppose, though, that this raid should not be listed as completely successful. It concluded our campaign.

An old arrastra on the Deerlodge National Forest. With the aid of a horse or mule, early-day miners dragged large stones in a circle in the pit, pulverizing the ore.

70,000 Mining Claims

Since the days of the early gold placer miners in Montana, mining claims have been filed continuously, sometimes at feverish speed after news of a strike leaked out, in the area comprising what are now Silver Bow, Deer Lodge, Powell, and Granite Counties. The United States Geological Survey in its research years ago termed the Philipsburg Quadrangle (in Granite County) the most highly mineralized one in the nation, indicating the great diversity of minerals there. In Butte, of course, copper had long since displaced gold, silver, manganese, and zinc in importance, but mining continued to dominate the lives of the people in this vast region.

As the Deerlodge National Forest lay in the heart of this intensively developed mining area and had mining claims scattered here and there within its borders, it became imperative that we acquire basic information regarding ownerships. Early in my supervisorship of this forest, having received instructions from Washington, DC, to make a comprehensive report on mining claims on the forest, I consulted the county records in the various counties and arrived at a rough estimate of over 70,000 mining claim filings on the forest. (And at that time, Granite County was not included.) I should have had a computer.

I noticed that the names of the recorded mining claims often tended to trace the history of the nation's development, which might make an interesting study in itself. For example, the Atlantic Cable Mine, a fabulous free-milling gold producer known generally as the Cable Mine, memorialized the laying of the first Atlantic cable. The Destroying Angel claim operated under the area on which the vice district of Butte was located. It was named after a semimilitary group of Mormons formed to defend their early settlements against encroachments of the Gentiles.

A mining claim is six hundred by fifteen hundred feet—fifteen hundred feet long along the vein and three hundred feet on either side of the vein. It comprises, for a full claim, an even twenty acres. However, many claims are fractional and many others have been located, abandoned, and then relocated many times. A patent to a claim under the mineral laws of that time conferred full title to the land, both surface rights and to the underlying minerals. The claimant to an unpatented claim could prohibit trespass, and these rights did not cancel merely because he was delinquent in his annual assessment work; he could still bring action for trespass against anyone occupying the claim or removing any of its resources. The claim was, however, subject to relocation, to what is called claim jumping, an often exciting and unpopular practice.

Both the patented and the unpatented claims were, therefore, of concern to us, since we must try to avoid committing a trespass. For example, when we made timber sales, we had to make every effort to be sure we were not selling the timber off some valid claim. There might be a strong case made for reimbursement for the value of the timber, even though the claim were unpatented and the assessment work not current. That was always a headache with us, and we had occasional difficulty. The law regarding surface rights on mining claims has since been amended, reducing the complications in such cases.

I remember one sale we made for mining timbers. Shortly after the timber had been duly cut, a man represented by counsel came to my office and proved conclusively that he had a number of claims right in the middle of the sale from which the timber had been cut. We had advertised fully in the newspapers before making the sale. Subsequently we learned that the fellow had filed on the claims, owned by someone else who had permitted his assessment work to become delinquent, after seeing our advertisements in the papers. In other words, he had jumped the claims. Then he waited for the timber to be cut and promptly demanded payment for its value. When confronted by the evidence, the fellow showed a great desire for anonymity. We heard no more of his claim.

In a mining community some people were not above filing mining claims for possession of the land for reasons other than mining. One day, while on an inspection trip, the ranger and I saw a new mining location notice right by the forks of the road near Georgetown Lake. It was an ideal place to put up a saloon or tavern, which happened to be the business in which the claimant was engaged. We could find no discovery pit, and we doubted that he intended to mine for gold, silver, lead, or copper, so we promptly visited him and told him that if he had made a discovery there and was going to mine, we were happy to hear about it, and would lend him all our cooperation. But if he intended to put up a bar, he would have all of our opposition. He abandoned the claim. It was a typical case.

A timber sale in lodgepole pine on the Deerlodge National Forest.
—Photo by Thomas H. Gill, US Forest Service

Another problem was that of building access roads in this area. We needed to build a short new road up to a ranger station out of Basin—only a matter of three-quarters of a mile or so. We found a couple of newly located mining claims posted right over our fresh survey lines. Presumably we could have ultimately had the claims declared invalid for lack of a mineral discovery, but that might require several years of examinations, hearings, and legal action. Otherwise we could not build the road without paying for a right-of-way. The claimant had a full right to forbid our trespass.

We learned that it was a good idea, if we planned a road construction project, to keep the matter as quiet as possible until we were actually building and had possession of the right-of-way. Otherwise, some of those so-called "paper hangers" would go out there and post location notices, forcing us to pay for a right-of-way. In other words, our motto was, "Be sure you're right, then go ahead and tell people about it afterward."

Carlson charcoal pit in the Boulder District of the Deerlodge Forest, circa 1925–1930. Wood was stacked in a standing position and then covered with soil to limit the air flow. Charcoal was desirable in the smelting process because it burns hotter than wood and is easier to transport. —Photo by Thomas H. Gill, US Forest Service

Besides the sometimes exasperating task of threading through a maze of mining claims, we sometimes had other interesting problems connected with land ownership. By far the most significant contribution I made on the Deerlodge National Forest was the initiating, planning, and finally consummating a great land exchange with the Anaconda Mining Company. In this deal, the company took title to some 180,000 acres of national forest land around Anaconda that had been damaged or denuded by fumes from the smelter, giving in exchange undamaged lands owned by them with the forests farther west. It was a mutually profitable transaction. The company, by taking title to the forest land it had damaged, freed itself of possible payment of damages. The lands it gave in exchange were scattered tracts within the national forests that it could not log profitably. Nevertheless, the Forest Service could use these lands in connection with its adjacent land and avoid the complications that come from mixed ownership. With modern logging methods, the land the Forest Service acquired in this exchange has become very valuable.

My eight years in Butte were happy ones. Butte is an interesting community with friendly, warm-hearted people, loyal to the hilt to their city and to all Butte people. We were eventually transferred from there, but we left with sincere regret.

Aerial photo of Milltown Dam on the Clark Fork River upstream from Missoula in 1930. —Photo courtesy Library of Congress

Mapping Gets off the Ground

The job on the Deerlodge National Forest during my years there as supervisor was in part a public relations one, in addition to its protective and custodial aspects. There were not to exceed thirty people on the entire staff, and yearly receipts amounted to only $15,000 to $20,000. We had to plan, however, for the time that would inevitably come, as it has, when there would be a great demand for these resources in multiple use. Since this was a publicly owned business, the public had its part in this planning. Its needs and attitudes would finally determine the details of our plans, but first the public had to understand the problems and be given information so that its contribution could be an intelligent one.

In addition, with the Forest Service as a whole, we had to train our employees to a far greater extent than many agencies since we had many lines of work and activities for which there was no source of qualified people from whom to recruit. To this end we had to institute formal training courses. Generally, esprit de corps was high; people willingly accepted challenges and recognized the need for improvement and for mastery of new techniques.

An example of one of these unexpected demands was our initial experimentation with aerial photography as a method of mapping our isolated country. Incidentally, with some hesitation, I made the first proposal in the Forest Service for aerial photographic mapping. When I became supervisor, I requested it specifically for the Whitetail Park area. I had served with the Fourth Engineer Regiment of the Regular Army during World War I and had used aerial photographs and mosaics in France. This experience convinced me that there was a big potential for the use of aerial photography for mapping inaccessible country where ground mapping was time consuming, expensive, and difficult. I knew—I had used it to compile maps of country that was inaccessible because it was occupied by the German army. Also, I had mapped rough country the hard way.

We badly needed a map of that Whitetail Park country east of Butte, but there wasn't enough money in our allotments for a costly ground-mapping project. I discussed the problem with Jack Lynch, a veteran Butte pilot, who was then building an air service. He became interested at once. He knew the area and was willing to fly his plane on the job, even offering an exceedingly low price to get the idea started. We planned to use aerial mosaics and the simplest methods for compilation. It would not have measured up to present-day standards, but it would have served our purposes; and it would have worked. When this proposal and plan reached the regional office in 1926, the whole idea was regarded as visionary and met with opposition. There were, however, those who supported it. Howard Flint, Region One's first chief of the Division of Fire Control, was receptive to the idea, having been thinking along those lines himself. Jimmy Yule, my longtime friend in the Division of Engineering, was enthusiastic. But their support was not enough, so I never did get my request granted.

It was not until two years later, in 1928, that the Forest Service let its first contract for aerial photography to be used in mapping. Progress! Nick Mamer of the Mamer Flying Service of Spokane, Washington, got the flying contract. Howard Flint happily took the pictures, using a Fairchild camera that was borrowed from the army. Jimmy Yule developed the compilation methods, trained the photogrammetrists, and did such a magnificent job that the US Army copied much of his work as it extended its use of aerial mapping. Subsequently, Jimmy headed a unit during World War II compiling maps for the army from photographs flown from the war zones. Donald Barnett of a well-known local optical company became interested and developed and manufactured eye glasses with special stereoscopic lenses for the use of those compiling the maps. These brought out the relief in rough topography in a spectacular manner. Obviously, there was no source from which we could recruit people qualified as photogrammetrists. We appointed men with training as draftsmen and as topographic engineers and trained them.

Unlike many government agencies engaged in regulatory or protective work, our biggest job, primarily, was administration. Any public or private board of directors or administrative council knows how difficult it is to hire executives to meet its special requirements. It was particularly difficult for the government because its salary levels were markedly lower than those in private industry. Moreover, civil service requirements and restrictions excluded many, either because they could not meet the one or would not comply with the other. Finally, the Forest Service set up its own administrative management training program organized by Peter Keplinger out of the Washington, DC, office. Part of the program was conducted by correspondence, but there were also meetings, conferences,

Rangers training at the Priest Lake training camp, spring 1928. —Photos by J. R. H.

and seminars in which men were trained in the techniques of administration—scientific management, it was often called.

Naturally, I became involved in that program, first as a student when I was a young administrator and ultimately as one of those directing the training as chief of the Division of Personnel Management in Region Nine. Training was a function assigned to that division. Ultimately, we even published a book or pamphlet that has been widely used outside the Forest Service as well, although I imagine by this time it has been supplanted by other books. At least I hope so, for there should be progress and improvement.

The pamphlet was a sort of anthology of articles or chapters prepared by personnel and training men throughout the Forest Service. The chapter on "Direction" was my share. This constant emphasis on development of systematic approaches to administrative problems contributed in a great degree to the efficiency we obtained in the Forest Service.

Long before we evolved this program of training higher ranking administrators, we had found it expedient to conduct training camps for young men who were about to assume the responsibility of a ranger district. For three or four years, in company with W. W. White of the regional office, I assisted in conducting a ranger training camp near Priest River, Idaho. We simply took a group of fifteen or twenty young rangers to the experiment station buildings below Priest Lake, where for a month we lived together and went over every phase of the work of a district ranger, with special emphasis on his administrative responsibilities. Most of the men were already competent in the technical phases of the work. There were also fire training camps in the spring on most forests, training in control of tree diseases and insect control, and in many other activities.

At the time of the economic Depression of the 1930s, the Forest Service, which up to that time was a relatively small, static organization with a moderate staff, suddenly found itself having to take over greatly expanded activities. The federal government assigned to the Forest Service responsibilities in CCC, WPA, NIRA, NRA, and other work relief programs. The Division of Operation in the regional office had to expand its staff and form sections for specific functions to meet the emergency demands and expanded load of work. Possibly, in part, because of my training experience, I found myself head of the newly formed personnel section. Eventually, when a breathing spell came, I got down to the serious study of personnel management. How this new work related to my professional training as a forester, I would not know, unless it was that I was dealing primarily with foresters and that an interest in all living things constituted a basic relationship.

CCC—Best of the Alphabet Agencies

I was fascinated with the CCC program, which I considered the best of the alphabet agencies in the Franklin D. Roosevelt regime. It continued from 1933 to 1942. The CCC was one of the brightest spots of the Depression years. I believe it is generally agreed now that we managed to do a great deal of good for many of the boys assigned to these camps throughout the nation. There were a few hopeless ones. In the early days of the program, some of the big cities were guilty of using the CCC to get rid of their gangsters. In fact, in one camp we found several men over the top age of twenty-five who were graduates of the prison at Sing Sing. However, we promptly turned those loose.

The CCC program started like an avalanche—right now! We had no advance warning until we suddenly received instructions from Washington, hastily drafted and incomplete, that we were to participate in a new program called the Civilian Conservation Corps. The administrative setup was unique in that it was a sort of three-headed monster. The actual management of the camp was under the direction of the US Army, which furnished the camp commander and other officers as his assistants, including a doctor for each camp. The Army had complete authority and jurisdiction over the camp itself, including housing, mess, camp discipline, and recreation.

The work program was under the direction of various agencies, the Forest Service having the largest number of camps. The National Park Service, state conservation departments, and some other agencies also had camps. We furnished the foremen and technicians and took responsibility for the boys from the time they left camp in the morning until they returned at night. We provided the work projects, the overhead, a few "skilled workers," the equipment and supplies, and directed the boys on the work.

Finally, with a rather vaguely defined but very real set of functions, a CCC office was set up in Washington under the direction of a man by the name of Robert Fechner. Fortunately, he proved to be an able administrator. He could exercise full jurisdiction, if he chose to do so, over our selection of personnel. A part of the program involved political patronage, something that probably was necessary in order to get the support of Congress. Various congressmen had the privilege of assigning us lists of foremen from whom we selected our project overhead. We had the selection of most technical and professional personnel, appointing whom we saw fit, though they were ultimately put under civil service. However, the distinctions between the two groups were ill-defined and confusing. With a combination of firmness and understanding, Fechner adjudicated the inevitable conflicts and misunderstandings and made it possible for us to work. He helped define where the authority of the US Army left off and that of the work agency began.

However, it was a weird administrative setup, and as I said, it descended on us all at once. We had all too short a time to build an organization and prepare when the boys began to arrive, hundreds of them.

Portal at the Ruby CCC camp on the Beaverhead National Forest in Montana. —Photo courtesy US Forest Service

Perhaps the very impossibility of the administrative system contributed to its success. Both the Forest Service and the US Army were aghast at the possibilities for trouble and conflict and consequently worked the harder to cooperate. Fortunately, right at the outset we didn't have those patronage lists—we had a free hand to select personnel, and we had to work fast. There wasn't time to obtain political recommendations of personnel. Over 2,600 appointments passed through my personnel section in the Region One office in Missoula, Montana, in a period of approximately six weeks, including hiring for other alphabet organizations.

Initially, we staffed the camps in part with some of our proven career foresters as superintendents, men whose capabilities we knew. We'd pull a district ranger and break up our regular organization to get the camps going, leaving the assistant ranger in charge of the district. The situation was simply too complex for us to trust to strangers who knew nothing of the often rigid government fiscal procedures and limitations or of our organization. It was not until later that we were able to work in such men as logging superintendents, who were available for the asking with much of the logging industry shut down.

We needed one or two technical foremen per camp. When there was considerable forestry work to be done, we used graduate foresters. If a lot of engineering work was required, we assigned fully trained civil engineers and construction men. One could pick them at random. Men were out of employment all over the country. And finally, we needed practical foremen for such work as road construction, mechanics, skilled operators of heavy construction equipment, carpenters, and a host of other skilled tradesmen. Many of these were men who had formerly worked in temporary jobs for the Forest Service. Others were from the shut down logging camps, from now idle road contractors, and from nearby towns. There was no problem finding men; the problem was to organize them, assign them where needed, and to provide them with orderly plans so their work would accomplish something.

Along with that, of course, there was the problem of employing the additional stenographic, clerical, and other office and warehouse personnel for payrolling, records, purchase, and dispatch of supplies and equipment and other jobs essential to backing up the camps themselves. This gave us an opportunity to give employment to a very large number of excellent clerks, stenographers, and accountants who were out of work. For a time, every young lady in town was smiling hopefully at me as I walked to work—and it came twenty-five years too late for me. I was glad to have my young brother-in-law with me to accept the smiles and greetings. He was of an age to appreciate them.

Then there was that problem of patronage. The Forest Service had always been a solid civil service organization. Virtually no positions ever involved patronage, and virtually no promotion or advancement ever took place as a result of any political influence. We usually did not know the political affiliations even of our immediate associates. Now, for the first time so far as I can remember in the Forest Service, we found ourselves filling positions at the dictates of politicians. It was a new and unwelcome experience and we had to learn to live with it. Of course, whether it was good, bad, or indifferent depended to a considerable extent on the politician with whom one was dealing. I found, somewhat to my surprise, that many of the congressmen and senators were very conscientious in trying to give us capable men. Senator Burton K. Wheeler of Montana went to the extreme of getting a retired civil service examiner to examine and rate the applications of a group of engineers and made his recommendations for appointments to us on the basis of that examination. Obviously, we got just as good engineers as if they were taken directly from civil service lists.

As I said, most of our congressional advisors tried to give us good men. One congressman said to me, "You know, every time you get a man a job, you make five or six enemies because they didn't get it." However, there were a few who were careless with their recommendations, disregarding

Chopping wood at the Ruby CCC camp on the Beaverhead National Forest in Montana. —Photo courtesy US Forest Service

character as well as qualifications. We simply could not risk the lives of the CCC boys with unqualified foremen, nor put them under the influence of men of bad character. We were trying to help these boys, not harm them. While I still think the proper way to select personnel for the government service is by impartial examination without regard to political, religious, or ethnic background, I have to admit that patronage was not quite as bad as I had been led to believe in my strict bureaucratic upbringing.

It was a busy period; we were working night and day. Presently the program began to shake down. We learned from our errors, and results began to show up. I suppose those who were never engaged in the program may believe that all the boys were products of the slums, underprivileged, but this was not so. In the Depression of the 1930s, the economic distress reached into all classes. People who had been actually quite wealthy were suddenly bankrupt and destitute. Large numbers of solid, middle-class people with good incomes were utterly helpless. It was an economic disaster. Therefore the boys, as well as the personnel we employed, came from all classes of society.

One day a quite prepossessing but tired looking man was directed to my office. He wanted a job, of course, but when I asked him what kind of a job, he said, "Oh anything—just a pick and shovel job, just work." I explained that I was not hiring laborers, that that was handled elsewhere. However, something about him made me question him further. It turned out that he had been the vice-president and general manager of a well-known construction firm that had gone bankrupt with the Depression. His salary had been $45,000 a year with a bonus, and he was a partner in the firm, but he didn't have a thing left. His home was in a west coast city. He had left what little money he had with his wife and children and had ridden freight trains back here to Missoula, where he had lived at one time and where he had gone to school.

He was unquestionably an unusually able and competent man and a qualified engineer, so I offered him a position as a road locator at $2,600 a year, which was quite a respectable salary in the early days of the Depression. The shock was almost too much for him—he nearly fainted away. I discovered he hadn't eaten for a day or two. The tragic end of the story is this. He went to work and was just as competent as one would expect. However, a few months later he was killed by a cave-in on a road construction project. His wife and children were brought to the funeral, and after the Forest Services the supervisor went to advise her as to the compensation she and the children would receive. But she and her children had disappeared, and we were never able to locate them.

Endless stories of the desperation and suffering of able, hard-working, prudent people during the Depression could be told. The human suffering was most visible to those in the personnel office, where people begged, sometimes in tears, for a job—any kind of work. Some of the work relief programs may have been inefficient—mere boondoggling—but to those of us who saw first hand the relief the work brought to despairing people, they seemed worthwhile. Moreover, the stories of workers leaning on their shovels were vastly exaggerated; the majority of the workers were willing. Inefficiency was more often the fault of the work planning by the responsible agency than due to worthlessness on the part of the worker.

It was all very well to put on paper that we had the work programs and the US Army had the camp management, but frequently it was pretty difficult to say where camp management left off and work programs began. Furthermore, both of us were supposed to participate in training activities. A lot of the training was on the job, but every camp offered classes for the boys also. There was an educational advisor under Army jurisdiction in each camp whose duty it was to organize and direct the educational activity. Both US Army and Forest Service personnel taught classes in the evening. We had hundreds, and perhaps thousands, of boys who completed their high school educations in these camps and received their diplomas from nearby high schools after suitable examination to assure that they had earned them.

There were some boys who were completely or functionally illiterate when they arrived at the camps. I remember one educational advisor who resorted to comic papers and found them the best material for teaching these illiterates to read. About the only texts for beginning reading were the "Run, Spot, run; see Spot run" type, which were not very stimulating material for boys eighteen or twenty years of age. Li'l Abner may not have been grammatical, but the comic strip cured several cases of illiteracy.

Perhaps one of the great benefits of the program was that in a given camp one would have boys from solid middle-class homes with good cultural backgrounds mixed with boys from the slums. It seemed to me (this is a purely subjective judgment) that the former did more to raise the level of the slum boys than any effect in the other direction. It wasn't a case of the slum boys pulling down the better-class boys; it seemed to work the other way. Also, I wouldn't know whether the college-trained men or the so-called practical men—mechanics and equipment operators—were the more successful in working with the boys. I can think of outstanding successes from both groups. It was, I think, very unfortunate that the camps were racially segregated, but the nation had not yet recognized and met the problem of civil rights.

One of the best foremen for the development of boys whom I appointed was an elderly elementary school teacher, John C. Orr, of Missoula, Montana. I hired him from a political list with fear and trembling, thinking he would have no practical ability to direct such work as road and trail or telephone line construction. He didn't do badly in that respect, but he was outstanding in his handling of boys. I visited his camp and found him one day with a crew of boys on the simple task of digging holes for telephone poles for a line to one of our lookouts. They were working in quite rocky ground.

As I approached the crew, Mr. Orr was sitting on a stump—just sitting there quietly watching the boys work. Nearby, the boys were working as though they had an overseeing slave driver behind them—the dirt and rocks were just flying. I couldn't understand what was taking place. We didn't want to work people to exhaustion, so I asked, "What's going on here? You're sitting on a stump, while those kids are working like they are scared or something."

"Oh," Mr. Orr answered, "they're conducting a research project."

Men stripping logs for a construction project in 1933. —Photo by K. D. Swan, US Forest Service

"Research project!" I said. "What do you mean? They're digging post holes."

"Yes," continued Mr. Orr, "but this group here is digging holes of small diameter." (I think he said twelve inches or so.) "That next group has it at eighteen inches, and the group over there is digging them still larger. The experiment is to find out, in this rocky ground, whether you can make more speed by digging them narrow and having less to excavate, or whether in these rocks, it's better to dig a bigger hole and be able to get the rocks out easier. Of course, we don't really care about that, but the boys are learning how to figure out the best way to do a job."

That was the way Mr. Orr handled those boys. Even on a menial task of that sort he was training the boys to think and to develop skills. As might be expected, his boys turned out well. His camp, in a period of one year, advanced about eighty boys to full-paying jobs. That was an outstanding record when one realizes that it took place when there was still widespread unemployment. It must be added, however, that his particular camp was filled with an especially desirable group of boys, boys from farms and small towns in the Dakotas and western Nebraska—boys from solid homes, who were used to working.

The real problem in the CCC was to find foremen, such as Mr. Orr, and superintendents who were both technically competent to handle the work projects and who also had that rare faculty of being able to inspire and develop boys. Yet we did find a great many such men. There were a surprising number of them who developed a keen interest in helping the boys make the most of themselves.

In Region Nine, which comprised nine states including Missouri, the Ohio Valley, and the Lake States, we had about four hundred camps, with a total enrollment of between fifty and sixty thousand enrollees. (This great number is understandable when one realizes that they drew from an area that had about sixty percent of the nation's population, according to the census of that time.) I visited a camp down on the Ohio River, made up of colored boys. I arrived shortly before noon and was escorted to lunch by the camp superintendent. There was a separate dining room for the camp overhead, the Forest Service men, and the army officers. When we entered the lunchroom, a handsome, clean, upstanding young African American in a waiter's jacket politely took my coat, hung it up, and as I went over to the table, held my chair for me. I never was served with greater skill in any good restaurant than I was there in that CCC camp. There were several other waiters doing the same thing for the other men present.

As we went out, the superintendent said, "Now I suppose you'll naturally have some criticism of all the style we're putting on and the nice service given to us foremen. However, before you criticize, I want to tell you that that's one of my training programs for these black boys. We've got to train them for jobs for which they'll be accepted. So, we're training waiters. We have a retired dining car steward who comes from Ironton. He trains the boys in the art of serving. We also have a retired black woman school teacher. She comes out just to give them training in grooming and manners. When she visits the camp, there's a separate table, and she has dinner with the boys, and afterwards reminds them of such little niceties as cleaning their fingernails—matters of that sort."

"Well," I said, "how is it working out? Have you actually succeeded in placing any boys?"

He dug out of his desk drawer a list. It listed several who, at the conclusion of their training, were working in good hotels. Others had followed their instructor and were serving in dining cars. It was a highly successful program. The unfortunate thing is that we had not yet reached the stage of integrating the camps and of hiring African American overhead. Probably that retired dining car steward should have been a full-time employee.

That particular camp superintendent was exceptional in more ways than one. For one thing, unlike many of our men, he preferred an African American camp, feeling it gave him more opportunity for service. He established outstanding records in many respects. I had known when I went down there that the camp had an outstanding safety record. There hadn't been a lost-time accident there for many months, but I didn't know how he achieved the record. The night I was there happened to be safety meeting night. Of course, I attended the meeting. It was conducted by the boys, themselves. I found they had a different group of safety supervisors each week. The boys in charge of safety for the current week brought before the meeting all cases of violations of good safety practices that they had observed.

The most interesting thing was that they had observed the overhead personnel as well as the boys. They had records of cases where foremen and even the superintendent had been guilty of some unsafe practices. For instance, one of the boys stated that when a shower started, one of the foremen ran for shelter carrying a double-bitted axe over his shoulder—a very dangerous practice. The boy reported it in a genial way, laughing a little as he said, "I think Mr. Jones had better get wet than take a chance of hitting himself in the head with an axe."

The incident may seem like a trivial matter. However, these were boys from nearby cities like Cincinnati, who had arrived at the camp undernourished, discouraged, and downtrodden. They were now men, showing initiative and dealing with the staff members with self-respect.

The program made a substantial contribution to social conditions in the area.

Perhaps the outstanding tribute to the success of the CCC program was contained in a letter I received from a successful professional man in the Midwest. He explained that his wife's family had been rendered virtually destitute by the Depression and that their oldest son—his wife's brother—had enrolled in the CCC. He told how the boy returned home a man, confident and filled with ambition. With his brother-in-law's financial help, he was now in college and making a good record. He asked if it would be possible for his wife's younger brother, also, to get into the CCC. He stated that the family was now self-supporting, so the boy did not qualify for regular acceptance, but he said that he, personally, would guarantee payment of all costs for the boy. They hoped he could duplicate his brother's experience. The final test of a product comes when people are willing to buy it, and, in the eyes of this family at least, our program met the test.

LEAVENING THE OZARKS

When the national forests were established, they were made up of public domain lands and, accordingly, were in the western part of the nation. Even there, there was inescapably a lot of land for which patent had been issued that was included within the forest boundaries. Many of these tracts were relatively valueless to their owners, sometimes because they had been logged off, sometimes because a homestead had not proved profitable and the people had moved out. These patented tracts often were obstructions in the administration of the forests. Therefore, the General Land Exchange Act was passed in 1922, which permitted the Forest Service to acquire such tracts, giving in exchange other land or timber of equal value. This permitted consolidation of holdings.

Another law, the Weeks Act of 1911, permitted purchase of lands for national forests, such lands for the protection of the headwaters of navigable streams. Since most land lies on the drainage of some navigable stream, this made possible purchase of land throughout the eastern half of the country where there were no unappropriated lands available for national forests. Progress in acquiring land and establishing national forests was slow up to the time of the Depression, then appropriations for acquiring land were increased. States and counties throughout the Lake States, the Ohio and Mississippi Valleys, and in the south held vast acreages of cut-over timber lands that had become tax delinquent. The states and counties were anxious to sell these lands to the federal government for low prices and thus recover at least a part of the delinquent taxes. Likewise, private owners of these lands often found them to be unprofitable and even liabilities during the Depression and were anxious to sell. Land acquisition became a major activity.

When it came to the acquiring of these lands, either by exchange or purchase, the Attorney General's office in Washington, DC, was exceedingly insistent on absolutely clear title. In fact, in the opinion of many of those involved in acquisition, that office often seemed unreasonably insistent. It required the clearing up of any conceivable flaw in the title back to the original patent. To us this appeared unnecessary, since in most states there were statutes of limitation that cleared up any question of possible challenges of title after a certain period of years. Naturally, there was a lot of complaint about this, and some weird stories were told. One story, undoubtedly apocryphal, illustrates the attitude toward the long search of titles that sometimes held up the transfer of or payment for land for as much as two or three years. This story involved a land purchase from an individual owner in Missouri. Repeatedly, the Forest Service was required to go back to the seller for additional information on the title. Finally, they were said to have received a letter from the seller that read something as follows:

"I have now taken this title back to the original patent from the United States of America. The United States acquired the land in the Louisiana Purchase from the King of France. He acquired the land in a war by conquest from Spain. Spain acquired the land by virtue of discovery by Christopher Columbus, a subject of King Ferdinand and Queen Isabella. They acquired the land, as I said, by discovery, and they had the right by grant of the Pope in Rome. The Pope was the vicar of Christ on earth; Christ was the Son of God, and God created the earth. That's as far as I can take my title."

When World War II broke out, the Forest Service had a competent administrative staff. It also, through its big land acquisition program, had developed a staff of men trained and skilled in appraising and purchasing land. The United States needed an area of over ninety thousand acres in the Missouri Ozarks, the tract that became Fort Leonard Wood, an enormous army training center. The job of purchasing this land was turned over to the Forest Service. In fact, we got our instructions by telephone, and vague instructions they were. They gave us the general exterior boundaries and told us "to get down there and buy that land as fast as possible."

Ray Harmon, assistant regional forester in charge of the Division of Lands in Region Nine (and later in Region One) was the man who actually did the purchasing. Of course, the rest of us in the regional office backed him up with the necessary services. My job was to find a lot of men who were capable of doing the examination. We didn't have a large enough staff to rush the job through in the time given us. There were, here and there, if we could find them, people who had worked for us on land acquisition, and we managed to get a lot of them for this job. We had also to get a lot of additional abstractors to examine titles.

It was a terrific job, and its accomplishment in something like three months was a feat of which the entire Forest Service could be proud. Of course, the major credit was due to Ray Harmon, who was right on the ground in the worst of it.

There were few condemnations. By skilled negotiation, our people managed to purchase virtually all of the land without condemnation and without paying exorbitant prices. On definite instructions, appraisals were reasonably liberal. The land was acquired in winter and possession granted in early spring. This meant that these people with their little hill farms were thrown out just at the beginning of the growing season, and inescapably were going to lose a year's crop. What they produced was chiefly for their own use—a little corn and sorghum which, together with their few head of cows and hogs, made up their livelihood. The appraisals were sufficiently liberal to make it possible for them to relocate, and assistance was given by another government agency in finding other locations.

The wholesale expenditure of money by the government in buying land and relocating families so rapidly really shook the Ozark economy. The idea of multiplying by arithmetic or even geometrical progression caught on fast. As soon as Fort Leonard Wood was established, the various staff officers came to set up administration. Initially, the government could not provide housing, but those who could do so wanted to bring their wives with them and to establish homes near the post.

There was one case reported of an officer who found a cabin just outside the boundary of the post, owned by one of the mountaineers. It proved to be a neat little cabin, so he asked to rent it. The mountaineer said, "Yes, I'll rent it, but I ought to get a little more money than I been getting."

"Well, that's a reasonable demand," said the officer. "How much do you want?"

"Well," the owner replied, "I been gettin' six dollars a month. I think now I ought to get sixty."

Here, with a vengeance, the multiplying tables came into their own.

Altogether, it was a busy, worrisome time, especially for Ray, who carried the main burden. However, even for men like me, who were backing him up, there were problems. We had to start so fast that we did not even know the statutory limitations of the legislation under which we were working. For example, I did not know whether I was authorized to employ people outside the civil service, people who had passed no examination and were not certified to me by the Civil Service Commission. Normally, one took no chances on a matter of that sort, for a government employee must take personal financial responsibility for the correctness and legality of his commitments. However, in the face of the war emergency, I felt fully justified in going ahead.

Perhaps I may inject a comment here on government procedure. It is fairly safe in the government service to make a great big mistake if it is done in good faith and not for any personal gain or profit. If one is going to make a mistake, it should not be less than a $50,000 one, and preferably $100,000 to $250,000. A mistake of that magnitude, made in good faith and in an honest effort to serve the government interest, will always be taken care of by a relief bill passed by Congress. The serious thing is to make a $100 mistake. If one did that, he would unquestionably and unhesitatingly be required to contribute a check for $100 to the US Treasury. So I was not seriously worried. In hiring such a large group of people, constituting a sizeable payroll, I was well into the safety zone. If I had been in serious error, I would confidently expect that I would be rescued by a relief bill.

Even before the tremendous upheaval created by World War II activities, the economic effects of the Forest Service going into the Ozarks, into Little Egypt in southern Illinois, and to the old worn-out coal fields of southern Ohio and Indiana, were striking. There was one county in Missouri in which the economists made a study. They reported that in this county before the war and in the middle of the Depression, the average income for rural families was $60 a year. The average family consisted of five—a man, wife, and three children. Twelve dollars per capita per year!

The strange thing is that their incomes were little if any lower during the Depression than they had been before. They never had had much cash income, and they didn't even realize they were poor. They had lived in the hills for so many generations that they really didn't have many wants. They raised what they had to eat, made all their own clothes, and even their shoes, if they didn't go barefoot. They were not bad people, nor were they lazy or shiftless. They were merely a people who had been stranded in an economic backwater.

Many of the very isolated Ozarkers were quite primitive, and they had their own social customs, which presented us a problem when we assigned young college men down there. We had to teach them some of the customs to observe. For example, in some areas, it was not proper for them to approach a home unless they were sure the man was home. If they were in doubt, the proper thing was to go up in the vicinity of the cabin and shout loudly. If the man were home, he would come out. If not, he might resent a call on his wife and children when he was away from home, and it was well not to incur his resentment.

In contrast to the average family income of $60 a year, the government's approved salary scale required us to pay a lookout $120 a month. As we put up lookout cabins, we hired local men as lookouts. Since they knew the country well, they were excellent lookouts and most of them were faithful to the job. However, some of them, after working a month, didn't see much point in working another month. They already had received two years' income. Why go on acquiring wealth when you already have it? That was a passing phase and didn't last long.

Gradually, something took place. We had comfortable little cabins at the lookouts with decent outhouses and cisterns for water—better housing than most of the mountain people had ever had. Furthermore, the men had money in their pockets. They could go to town and make purchases, buy clothes for their wives and children. After a while, Mrs. Lookout's sister was inclined to press her husband; why couldn't she have some of the things her sister and her children had? She'd like to be prosperous, too. It was a little like putting yeast in a bread sponge or, to speak in woods terms, putting sourdough starter in a batter. It leavened the whole Ozarks.

Another thing contributed to the same process. We had had a problem in this area finding constructive work for the CCC camps preceding World War II. When fully manned, we had two hundred enrollees in each camp; probably they averaged around one hundred and eighty boys besides the overhead. There was a limit to the amount of forest improvement work that one could do. Until we got sufficient nurseries, we could plant only limited areas. And in winter, stand improvement and planting were both almost impossible. However, we did need ranger stations.

The ranger stations we built in the Ozarks were in sharp contrast to those of the early days in the West, and at times we were quite severely criticized for some of them because of the amount of labor put in. With the need for training the CCC boys in useful trades, the supervisors built some quite elaborate dwellings. Some of them had rooms panelled in knotty pine or other woods. They had fireplaces. They were veneered with excellent building stone, quarried by the boys themselves. We had to be careful, of course, about competing with private industry, but there really was no private building industry around there, and the boys needed practical experience to complete their training. Using the building of ranger stations, warehouses, and office buildings at the ranger headquarters as a workshop, we turned out a lot of boys who were qualified quarry men, stone masons, brick masons, carpenters, plumbers, and even electricians. When the end results were appraised, I think we accomplished a great deal.

There is no denying, of course, that the Forest Service wanted these nice stations. Ozark assignments were rather unpopular with a majority of our rangers, particularly if they had children to educate. Stations of this quality were a help in holding men there. Furthermore, we gave the people in those little southern Missouri communities an idea of a standard of housing that simply hadn't occurred to them. The homes in the entire area began to improve. The people developed some wants that made them strive to develop their resources. It was a stimulus that tended to motivate people and was definitely of benefit to the entire community.

Wood, Woodsmen, and War

Because we had no outside source of men trained in our technologies, training became an ever more important Forest Service activity. We continued to have ranger training schools, fire training schools for temporary firefighters, and for all sorts of other training activities. In Eagle River, in northern Wisconsin, we seemed to be conducting a series of such sessions almost continuously. We facetiously referred to the staff as the three horsemen.

In the late 1930s, however, we enlisted the aid of the University of Michigan to augment our training of junior administrators. While my arrangements were with the School of Forestry and Conservation under Dean Sam Dana, he selected J. C. S. Benson of the School of Political Science to conduct our first intensive month's course in administrative management. The University of Michigan men did a magnificent job of conducting the course. We continued the program each year until World War II interrupted it. Our men willingly took leave and paid their own expenses for a month at Ann Arbor, Michigan, to learn the latest techniques and philosophies of their jobs. We could not legally pay these costs from government funds.

We encouraged our senior staff members to follow this continuing education program. Not only did they attend institutes and important meetings dealing with their specific fields of specialization, but we arranged with the University of Michigan for a series of evening lectures, seminars, and discussions conducted by a distinguished authority on public administration.

At the outset of World War II, when our nation had to get its military machine to expanding and moving fast, we all helped in a variety of ways. I was borrowed from the Forest Service to help in the instruction of personnel officers for the war agencies at two civil service training schools. One was set up in Chicago and later another was conducted in Greenwich

Village, much to the delight of my friends. At Chicago, a question was asked: How could they—the war agencies—carry on training when the law didn't authorize them to do so and there were no funds appropriated for that purpose? Arthur Flemming of the Civil Service Commission tossed the ball to me. He said, "We'll let Mr. Taylor answer that one if it can be answered, because the Forest Service has carried on training all the time, probably more than any other agency in the federal service."

So I found myself having to explain our rationalization of our training schools. I told them that the law by implication establishes our authority to supervise our employees. Supervision is inherent in conducting any operation.

"Now," I said, "how does one supervise an employee? He tells him what is to be done and tries to arouse interest in the work. He then tells him how to do it, for naturally any agency wants the work performed by the methods it has selected and to the standards it regards as appropriate. The supervisor may even demonstrate how to perform an operation. Finally, one has the employee perform the task under supervision and suggests improvement and points out errors. It's merely a coincidence," I explained, "that these supervisory steps happen to be comparable to the Try-Tell-Teach Test formula used in vocational training. We are supervising our employees and we are acting within the law."

Then came another question: how could we conduct this training without specific funds appropriated for it? That, too, could readily be explained. If one has to carry on a line of work, it is proper to pay supervisory costs from the funds appropriated for that work. This is one way to assure that training does not become an end in itself. We did not carry on training for the sake of training, but only as it was necessary to getting the job done efficiently to acceptable standards.

Certainly our constant emphasis on training contributed to the fact that during the war the government drew heavily on the Forest Service for jobs completely outside the realm of forestry. For example, we had to develop the propagation of guayule for rubber, with plantations and extraction plants mostly in California, though I believe there was one in Arizona. And we had the peculiar job, for foresters, of raising Russian dandelion. Can you imagine foresters raising great fields of dandelions? There was an Asiatic species of dandelion that for years had been used as a source of rubber in Russia. With their plentiful supply of cheap labor, they could do it. We raised the dandelions successfully and produced about sixty pounds of latex to the acre, which was made into tires for testing. It proved to be a high quality latex, superior to that obtained from guayule, but our labor cost was prohibitive.

Like any patriotic organization, we attempted to avoid asking deferment from the draft of men of military age. Also, we were willing to make men available to the defense and armed services agencies. We were not guilty of evading civil service regulations against raiding other agencies for personnel. However, it was done to us. We had plenty of trouble in that respect.

I imagine that to most people it would appear that forestry had little relation to defense and the war effort, but foresters did have considerable to contribute. There was a great deal of work connected with supplying forest products for the armed services, particularly in certain specialized fields. Some woods were in very short supply. Our sources of supply were cut off, primarily by the Japanese invasions in the Pacific.

For instance, shuttle blocks for the weaving industry had been made from tropical woods, which were virtually shatter proof. The Forest Products Laboratory informed the textile manufacturers that our eastern flowering dogwood, and to some extent persimmon, were suitable for shuttle blocks. But neither of these species grows in solid stands; they are scattered, and our men had to locate supplies of them, almost by individual trees. Likewise, white ash makes the best timber for lifeboat oars, and white ash, too, doesn't grow in solid stands but is scattered through woodlots in the Midwest, primarily in the Ohio Valley. That had to be found. Walnut for gunstocks, birch suitable for plywood for airplane veneer, and a lot of other specialty woods were vitally needed. The volume was not great in any one of these, but it was large in relation to supply, and the amount of time and work that went into locating it was disproportionate to the volume.

It was in connection with this work that I pulled a real "boner." We had a young man, a native of southern Illinois, who was an expert on the hardwoods of the Ohio Valley. He had, after graduation from a Midwest forestry school and after working for us for a while, gone to the University of Idaho at Moscow, Idaho, to study for his master's degree. While there, he registered for the draft with the Moscow draft board. Now remember, that is in the western white pine logging area—white pine, ponderosa pine, and Douglas-fir are the principal species. There, a large sawmill may turn out one million board feet of lumber per shift—possibly more now.

We decided we had to ask for the deferment of this young man, who had returned to us, since he was so essential in locating the specialty woods needed by the armed services. Accordingly, I put in a request for deferment. In justifying the request, I stated that in the past three months this man had succeeded in locating over three hundred thousand feet of specialty woods. To the draft board in Moscow, three hundred thousand board feet sounded like about three hours' work. To them it sounded plain silly, and I utterly failed to convey to them what was involved or its importance. I suppose what I should have said was that he had located timber for about ten thousand oars and several hundred thousand shuttle blocks for the weaving industry that was producing uniforms, blankets, and what not for the armed forces. Then they might have listened to me, but as it was, they promptly drafted him. It took three men to replace him, and we were very short of men.

Then there was quinine, which had been obtained chiefly from the Dutch East Indies and was now in short supply. It was absolutely essential for the control of malaria, since effective synthetic drugs had not then been developed. Quinine originally came from Peru and was called Peruvian bark. However, the quinine industry had been transferred to Malaya, the East Indies, and to some extent, to the Philippines, where they raised the cinchona trees from seed brought from South America. The Japanese promptly cut off that supply, which caused consternation, to say the least. We were throwing hundreds of thousands of men into some of the worst malaria infested areas of the world with our quinine supply cut off. Of course, pharmacists immediately started work on development of synthetics, which later became very effective and probably are used to a greater extent than quinine today.

Meantime, a frantic cry went out to locate quinine. Cinchona is found primarily in the tropical forests of Colombia, Ecuador, Bolivia, and Peru, where it grows wild scattered through other species. There are also a few long neglected plantations in Guatamala, where an effort had once been made to start an industry. It was necessary to send men with some knowledge of tropical botany down there to locate it, not only to locate supplies but to get it into production as well. South America had literally gone out of the quinine business, but the local governments were willing to permit our men to locate the trees we needed. We curried the Forest Service for men who had had experience in the tropics and sent them.

These career men, sent on a job that was vital to the war effort, went to South America as civilians, not as soldiers. This status created one of the injustices that occur in war. Most of them suffered great hardship working through the jungles. Some were injured in the inevitable accidents inherent in the job or contracted tropical diseases. Yet they returned with absolutely no legal rights for job placement or preferment such as existed for those in the armed services. I do not believe it would be possible to formulate laws or regulations to obviate all of such injustices. We tried to give recognition to the merits of their cases, and probably they did not suffer unduly in career advancement, but some did have to take less desirable assignments, initially at least, because they lacked veteran's preference. This points out another problem in trying to deal fairly with men.

After World War II ended, we set a standard for our Division of Personnel Management that all returning veterans should always receive a cordial welcome and careful counselling regardless of how heavy the work load was. That applied to our nonveteran "specialty hunters" also. They were given first priority. We were determined never to take advantage of the legal provision that we were obliged to reemploy a returning veteran only if he was physically able to discharge the duties of the job he held when he left for the armed service. This meant that we had to find jobs for partially disabled men that were within their physical capabilities, sometimes retraining them for work new to them. At times it was a discouraging and frustrating business, and we wondered if it was worth the effort. Some of the men went out of their way later to express their appreciation of what we had done.

Two very thoughtful, far-seeing regional foresters made my job of chief of personnel management in the regional office in Milwaukee one of increasing growth and interest. The first was Lyle F. Watts, formerly head of the Northern Rocky Mountain Forest Experiment Station in Missoula, Montana, and ultimately chief forester of the United States. When he left us to become regional forester of Region Six at Portland, Oregon, he was succeeded by Jay H. Price. These two and their associate, Stanley F. Wilson, were all men of deep human understanding who made my job of working with people a pleasure.

While in Region Nine of the US Forest Service, I was privileged to collaborate with Aldo Leopold on a three-member oral examining board for the Wisconsin Civil Service Commission. The examination resulted in the selection of Ernie Swift as the director of the conservation department of that state. Both Leopold and Swift became eminent as conservationists and authors and achieved national recognition.

At that time, we had more than five thousand people on the rolls in an area covering Minnesota, Michigan, Wisconsin, Ohio, Indiana, Illinois, Missouri, Iowa, and North Dakota. In North Dakota there was little activity, chiefly one nursery and some shelterbelt planting. In Iowa, our work was chiefly extension work in farm woodlots. In all the other states, there were farm woodlots. In addition, there were two forest experiment stations within the region, and the division of which I was chief handled their personnel work as well as that of the region.

Raphael Zon, director of the Lake States Forest Experiment Station in St. Paul, Minnesota, was a profound scholar, a distinguished man who had an unusually interesting personal history. He was a Jew, born in Russia, whose family fled in successive stages from Russia to Poland, Belgium, England, and ultimately the United States, where he became a distinguished scientist. Zon spoke Hebrew (and no doubt Yiddish), Russian, German, and French fluently, but his English accent was atrocious. He was a rugged individualist who said exactly what he chose to say regardless of the occasion. He was never known to bow down to any person of distinction. Moreover, he was delighted when subordinates questioned data or policies. He encouraged them to argue with him. He had no use for subservient subordinates.

One day he was conducting a tour through an area of his field experiments for a congressional committee that was investigating Forest Service activities. He said, "Now, gentlemen, we are coming to two very interesting plots. On your right there is a Republican plot and on your left is a Democratic plot."

"But Doctor," said one of the congressmen, "there is something wrong. What do you mean by Republican and Democratic plots?"

"Well," answered Dr. Zon, "in the Democratic plot on your left, we have cut out all of the big trees so that all of the little trees can grow bigger, while in the Republican plot on your right, we have cut out all the little trees so that the big trees can grow still bigger."

For some reason, Dr. Zon could get away with such remarks.

HOME TO OUR MOUNTAINS

At last, heeding a nostalgia that grew stronger by the month, I headed west with my family across this land of kettle moraines, its lakes and hills and rolling prairies—back to our mountains. People in the Lake States still consider themselves residents of the Northwest, so we must add a name to be more specific about our location—the Pacific Northwest. We refer to our Montana as the land of shining mountains or the Big Sky Country. No matter, all returning voyagers remember the thrill that comes with the sighting of those first snowy peaks again. Home!

For years now I have eagerly revisited my favorite mountain and wilderness areas, noting the changes, some good and some bad. Now, an amazing network of roads spreads over the country like a giant spiderweb. Add to this improved and innovated means of transportation such as campers, jeeps, trail bikes, boats, and snowmobiles, and one can almost see the wilderness retreating. Logging roads have opened up so much formerly remote and almost inaccessible country, luring hunters from all over the nation in the fall, that it is actually rather hazardous to explore the woods at that season. But the migratory flocks of birds still fly over us in spring and fall—the snow geese, the Canadian honkers, the swans, and myriads of ducks. The bald eagles return as before to feast on spawning salmon in Glacier National Park.

Incidentally, a treasured homecoming gift from a longtime friend was a forty-year-old sourdough starter, an essential item for early-day pioneers and woodsmen. There are starters and starters, but this is the best one I have ever had. It has functioned and renewed itself for twenty-three years more, putting it in the over sixty class. It has been fun to revert to pioneer food. I have shared it with discriminating friends far and wide.

In summer I have checked and rechecked my early impressions of the mountains and valleys of the Bitterroot, Flathead, Blackfoot, Big Hole, and Clark Fork drainages, the Bob Marshall Wilderness, and Glacier

National Park. On backpacking and trail ride trips I have penetrated again the exhilarating backcountry. It was a boost to my self-esteem to find I was still in the one-blanket class on horseback trips. Yes, Sappho Creek is still Sappho Creek, and most of the early names of topographic features remain, but they are just identifications, lacking historical significance for more recent residents of Montana. Spurred on by my hunger for mountains, I have journeyed beyond my old stomping grounds, far north into the Canadian Rockies and on to Alaska and the Yukon.

On one of these backpacking trips with my family and grandchildren, as my wife and I were plodding up a steep, alpine trail in Glacier National Park, I was stopped by a passing park summer naturalist who asked me, "Why do you do this? Why do you come into this rugged country?" For me there was only one obvious reply: "Because I prefer the works of God to those of man." He whipped out a little notebook and recorded my remark. I wondered about that until I saw it published in an outdoors magazine years later. Did he approve or disapprove?

On the credit side are the exhilarating vistas of this great country, but on the negative ledger, causing me great sadness, no, anger, are the ravaged hillsides created by bulldozer-happy highway engineers and the vast areas of inundated land—the drowning of formerly green valleys, gulches, and canyons so army engineers could send electricity out of our state to Pacific Coast cities. To use a local expression, the organization seems to be "dam happy." My fishing day is ruined when I come around a bend of the stream to discover engineers' stakes, evidence of plotting to change Nature again. All in all it adds up to the raping of the land. And what of the native trout succumbing to those lethal doses of nitrogen below the dams? And is there any fate but starvation for the deer population pushed out of its longtime habitat? Progress? "I often wonder what the vintners buy one-half so precious as the stuff they sell."*

My deepest concern is directed at the determination of the Army Engineers to fill all of our mountain valleys with reservoirs, to milk every drop of energy from our now free-flowing streams. The question is one of values. The engineers feel that if a damming project holds promise of returning a fair dollar profit on its cost, that is all that is necessary to justify it. Even here their computations are open to challenge. Their original cost estimate is frequently far below the final cost, and when the estimated dollar-return is low, they simply charge a larger part of the total cost to flood control. But dollars are not a true measure of cost. How can one

---

*From poem attributed to Persian poet Omar Khayyám, translated by Edward Fitzgerald.

measure and express the cost of destruction of environment, of drowned valleys, animals deprived of their living space, of destroyed rivers? Neither are dollars a true expression of the benefits derived from these hydroelectric projects if one tries to measure their contribution to the happiness of people. Will the vulgar, garish advertising signs of Seattle, Tacoma, and Portland, the aluminum plants and the factories devoted to producing beer cans destined to litter our highways, and other profligate uses of electricity actually yield as much return in happiness and richness of life as would compensate for what has been destroyed?

Unquestionably we face an energy crisis. Ultimately we may have to harness all of our rivers to provide for truly essential needs. However, we can well postpone the evil day. Possibly in the meantime our scientists may have developed other energy sources. Even if they do not do so, we cannot go on indefinitely demanding more and still more energy for more and more wasteful uses for an ever-growing population. Finally there must be a limit if we are to preserve a way of life worth living, if we are to retain any part of the America we love.

Last year on a postman's holiday and sentimental journey, my wife and I took a look at Europe's Alps, following the majestic ranges from France's Mont Blanc to Germany's Zugspitze. We found them spectacular as we had anticipated, but under control of man with their tramways and cable cars. Perhaps the burgeoning population of those countries made this inevitable. We hope nothing like it occurs in Montana.

I noted the forestry practices through that area and found the controversial clear-cutting to be the accepted system of silviculture. Even near widely publicized tourist tracts, as in the Sound of Music country at Salzburg, Austria, I saw great expanses of clear-cut forest. No one seemed to object. The forests were utilized intensively down to the smallest branches. Problems are beginning to develop, however. As the residents of little villages and boroughs have a longtime vested right to the brush, twigs, and mast left after the timber is cut, there is little vegetation left to supply essential nutrients to the soil. This robbing of the earth is creating a complicated problem difficult to solve.

As the forestry practices in our country were built on the German forestry concepts, as taught by early-day forestry leaders, including Fernow, Roth, and Schenck, perhaps we can understand the transfer of the clear-cutting idea. On that subject—clear-cutting—we now have vehement groups both of the forestry profession and of the general public discussing and arguing the theory with both sides becoming more and more polarized in their positions.

151

We are not alone in our problems. In the British Isles all natural resources are carefully guarded. Going from the lakes country of England where a bulge of land 1,000 feet in elevation qualifies as a mountain, I was especially interested in reforestation being done in the highlands of Scotland near Edinburgh. Incidentally, a mountain, or *ben* as it is called there, has to be 2,000 feet high to qualify. Centuries of the old Caledonian forests were burned and the Scottish Highlands denuded, partly, no doubt, in the wars with the highlanders and partly to help exterminate wolves, which preyed on sheep. Over fifty-three years before, as a student at the University of Edinburgh, I had seen the early efforts to reforest the heath lands. Now as our bus climbed into the Moffat Hills we saw thousands of acres of thrifty forests—Scots pine, spruce, and larch. Vast areas through the Trossachs and over the bens and moors around Loch Katrine and Loch Lomond which I remembered as wholly given over to heather, whins, and gorse were now solid stands of timber. But I became aware that the forester is not always regarded as a public benefactor. Our bus driver bemoaned the fact that the foresters were ruining the beauty of the country with monotonous stands of forest. He preferred the braes and bens covered with purple heather and yellow broom. Others criticized the forest plantings because they conflicted with game—red deer, roe deer, and grouse. At home, in Montana, the foresters were also under attack, but there it was for clear-cutting some old, decadent stands and opening up the country.

The mountains of our Pacific Northwest passed my comparison test with Europe's Alps. All are beautiful, but these Rockies are my mountains, soul satisfying, evoking a reverence produced by no others.

I watch the terrific struggle going on, with wheels within wheels, trying to save our dwindling natural resources—our forests, our minerals, our wildlife. I note the greed for these things, the empire building and grasping for material wealth; but I observe also a change in a sense of values, a return to idealism on the part of our young people, our future leaders. I was impressed recently with the address of one of our local businessmen to the Missoula Kiwanis Club in which he declared with great sincerity that "we share our lives."

At the turn of the century young men were idealistic. It is refreshing to have this trait reappearing in this generation. They are developing an awareness that they must protect our heritage for themselves and for their children and grandchildren. Hopefully, the trend will continue into the future. More and more the public is becoming informed on matters of conservation. This is good. A fully informed public is our salvation. Never underestimate its potential.

I have devoted my life to conserving our natural resources here and in the Midwest. I am glad I was a forester. Even now this background brings me greater enjoyment in our mountains. Each season creates its rewards along with its concerns. The buttercups and pasque flowers, or crocuses, are in bloom now. Will this be another great beargrass year?

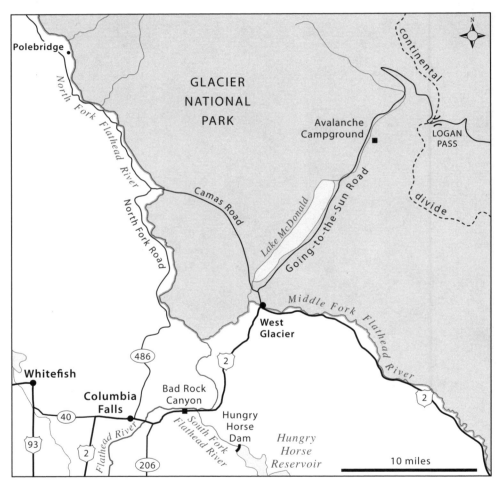

The journey to Logan Pass in Glacier National Park.

To Logan Pass

A few years ago while on a trip in the Bob Marshall Wilderness, our party made a day's ride on horseback, which was something of a record. In one day we rode some two thousand miles north, botanically speaking, that is. In fact, our whole family has made an equally long journey in an hour or so in an automobile when driving from our valley home up some mountain pass. In both cases, we started out among the familiar flowers and trees of our latitude and literally paid a visit to the arctic, or at least to the subarctic.

Thousands of tourists make the same trip every summer in Glacier National Park, but few of them realize it. Perhaps there should be signs along the road. For example, when one reaches the switchback on Going-to-the-Sun Road, he might encounter a sign reading, "You are now in the botanical latitude of Banff." On top of Logan Pass there would be another sign reading, "You are now at the northern (or upper) limit of tree growth in the arctic." Then, perhaps, a few people would realize that they have the privilege of seeing the authentic vegetation of the northland—that the flowers and trees they are seeing are much the same as they would observe while driving up the Alaska Highway to the Yukon.

One of the commonest questions asked by tourists is how high one has to go to reach timberline. What they do not realize is that there is a gradual transition as one ascends, and that timberline marks a zone of plant life composed of associations of subarctic and arctic plants left behind after the last glacial age. People usually seem to think of timberline as a sudden line of demarcation—that up to that point one is in essentially the same forest as in the lowlands, and that suddenly it terminates as though someone commanded, "Thus far and no farther." Instead, timberline reveals the great drama of the living plant striving for survival, evidence of the marvelous adaptability and tenacity of living things.

What I have been referring to, of course, is the climbing of our mountains and the effect of altitude on vegetation. I hope I can help you comprehend better what you will see when you ride up to Logan Pass in Glacier National Park. In effect, let's take the trip, only this time I shall not have to watch my driving but can point out something more than the distant view. I write this botanical tour for my grandchildren and for those who are humbled by our natural environment.

Before we start, it will be well to risk boring ourselves and to assimilate a few basic concepts; then we shall know what to look for. Plant ecologists, those who study the relation of plants to their environment, have a rough rule of thumb. They say that one thousand feet of altitude is the rough equivalent of about six hundred miles of latitude in changing temperature and in consequent effect on vegetation. Now, Logan Pass is some three or four thousand feet higher than Kalispell, so according to the rule, the vegetation at the top should resemble that found more than two thousand miles farther north. That would put one well up into the barren lands of Yukon Territory. There one would expect to find himself at the northern limit of tree growth and in a vegetation composed of sedges, mosses, stunted shrubs, and a myriad of quick-growing flowering plants. Well, that's just what we find on top of the pass. Moreover, many of the species are identical. At one time, while working around timberline in the Anaconda Range west of Anaconda, Montana, I amused myself identifying the plants growing there. That winter I read in the *Journal of Ecology* a list of plants found at the foot of a receding glacier in Alaska near sea level. About fifty percent of the species listed were those I had collected here in Montana on tops of high mountains.

The changes in composition of vegetative cover as one ascends a mountain are, of course, due primarily to lower temperature and to attendant circumstances of storm and snow. But this does not fully account for the presence of arctic (or alpine) species here in Montana. If one were suddenly to raise a section of Dakota prairie to an elevation of 10,000 feet, it would be just as cold there as at the tops of the peaks in Glacier National Park, but there would be no heather, no glacier lilies, nor delicate alpine penstemons. For these plants, together with the spring beauties, queen's cup, saxifrage, and alpine sedges don't have seeds equipped for long distance travel. They are not like the ubiquitous dandelion with its sailing seeds and marvelous adaptability, and even if they had greater facility for dispersing their seeds, they never could survive to cross the dry, hot prairies of eastern Montana. Similarly we cannot suppose that our mountains arose from a plain and miraculously became clothed in the wealth of summer flowers we see there. Neither is it reasonable to assume they always have been here, isolated from their northern relatives, for over the ages the steady processes of evolution would have been operating and there would be different species, related remotely if at all, than those of Alaska and the shores of the Arctic Ocean.

It is easy to account for the presence of arctic plants here. At one time the arctic had a temperate climate. Around the whole northern hemisphere extended a great circumpolar forest, probably with similar species throughout and continuous around the earth. Then came the successions of glacial ages. As the great ice masses pushed slowly south, foot by foot, over a period of thousands of years, they modified the climate in front of them. The circumpolar forest was pushed ahead of them while more delicate temperate and semitropical species retreated still farther south. Or as happened in some localities, if they encountered oceans or other inhospitable barriers they were exterminated. As each glacial age subsided and the ice withdrew north, the vegetation followed, but as it went it left islands of its components stranded on cool mountaintops. What we see on Logan Pass and other high peaks in the Northern Rockies are abandoned colonies of the latest glacial age. When the tongue of the Columbia Ice Cap was carving out Flathead Lake and pushing up the moraine south of Polson, the heather and saxifrage were growing in what are now wheat fields around Charlo, Ronan, and Pablo. As the ice crept north, they followed, while wheat grass, fescue, rabbitbrush, sage, and harebells occupied their abandoned homes. But up on the mountain they found a home suited to their needs and remained.

It is interesting to speculate on what will become of these isolated, abandoned colonies. Many, if not most of them, are stranded a thousand miles or more from any continuous population of their fellows. Below them lies an area as alien and inhospitable to them as the ocean is to a herd of buffalo. Most geologists believe we are now in the latter stages of the latest glacial age and that the ice is still receding. If so, the flowers on Logan Pass are doomed unless one of two things happens, either that the glaciers decide to return or that the great folding and faulting that pushed up these mountains raises them still higher faster than the processes of erosion can wear them down.

What we shall see on this trip are not "the everlasting hills"; there are no such things. Rather we shall see evidence of the everlasting processes of change, of the struggle of life for a place in the sun and for survival and propagation of its kind. Our mountains are not static; they are almost terrifyingly dynamic. As we start, let us remember that just a brief geological minute ago things were not as they now are, and another brief ecological moment from now they will be changed, never to be the same again.

Where we start out at Kalispell, we are nearly at the lower timberline. There are two timberlines in Montana, an upper and a lower. Just as the upper one is determined primarily by temperature, the lower is set by moisture. Down on the Flathead Indian Reservation it is practically treeless. However, here in the old lakebed that is the Flathead Valley, the fine silt holds moisture well and there are still scattered stands of yellow pines and Douglas-firs in the lower valley and solid forest farther north. When I first saw the valley in 1903 the real forest started only four miles or so north of town. It was magnificent—great towering larches, firs, and pines, two hundred feet tall. Now only farms and cut-over lands with scattered areas of second growth remain. Around Kalispell it was undoubtedly semiprairie when the first white men arrived. Take a good look at the yellow pines, properly named *Pinus ponderosa* or the ponderous pine. They are not fond of the arctic climate and you will be leaving them soon after you start to climb the mountains. They are rugged individualists and abhor crowds, a trait with which I am in sympathy. Botanists say they are intolerant, that is, they cannot endure shade. Thus, due both to their aversion to continuous cold and to their fondness for sunlight they will not accompany us far up in the mountains.

Look over to the right along that stream just a few miles north of Kalispell and note that silvery-leafed shrub. Probably you thought it was one of the many willows that crowd the banks of the stream. Actually it is an interesting survivor of the circumpolar forest, the silverberry. At one time its ancestors must have been a part of the continuous forest across what is now Bering Strait and on into northern Europe. Then the glacial age broke the continuity of the forest and a portion of the species traveled south in North America while the rest of the family remained in Siberia. The Asiatic cousins were more prosperous and successful, and they survive as the Russian olive, widely planted for shelterbelts and for ornamentals because of its silvery leaves. Our silverberry, though it has the same silvery leaves and a blossom of almost overpowering fragrance, grows in a spraddled out, unbeautiful form so it cannot compete with its Siberian cousin for our favor. Incidentally, its generic name is *Elaeagnus*, from the Greek, meaning "sacred olive," though it is neither sacred nor an olive.

The vegetation continues much the same to the mouth of Bad Rock Canyon, and then the change of type is less a result of altitude and temperature than of soil and moisture. More summer showers sweep over the mountains and canyon, favoring trees, and the soil is more rocky with less of the fine silt, which builds a tight sod in which tree seeds find it difficult to compete. Just such little factors determine the fate of a species.

Each has adapted itself to some particular environment. Each produces thousands of seeds and is ever taking advantage of any change in its surroundings, which may enable it to crowd out the others. The relatively dry but fine and fertile soil of the valley favored grasses and herbaceous plants. Here just a little more moisture, a different soil, and possibly a change of a degree or so in temperature is providing its chance to a whole new society. The forest becomes solid and almost unbroken up to the cliffs of the canyon wall, where mosses, alumroot, meadow rue, and others seize on every little crevice to establish a foothold.

On these cliffs we find another plant of which we shall have more to say later. Look for a small, scalelike incrustation on the rocks. Sometimes it is a mere spot of gray, hardly distinguishable from the rock itself. This is a lichen, the pioneer of the plant kingdom. This is the plant equivalent of Daniel Boone, or Lewis and Clark and the mountain men. It goes where no other plant dares. Shortly after the receding ice of the glacier lays bare a boulder or cliff, some bold lichen establishes itself. A beachhead has been established from which will spread the various forms of plant life. Unless the glacial ice or a lake comes to the rescue, the rock is doomed. It will require thousands of years, but the hordes of plants will never cease until, aided by wind, frost, and moisture, they have reduced the rock to soil suitable to your garden. But this battle with the rocks here in the canyon is just the mopping-up action; the forefront of the action is now far up on the mountaintops. This was the front line a few thousand years ago when the glacier was just backing up into the canyon and the present site of Kalispell was the bottom of the lake. We may as well wait until we get up around the front to study these skirmishers. There we shall see them at their intrepid best, just as they are along the shores of the Arctic Ocean or where the ice cap of Greenland has just laid bare a rock.

Now we have passed Hungry Horse, where we encounter an example of how some plants are capable of making an ally out of an enemy. Here there have been in the past a series of forest fires. In places the tall, gray spires of fire-killed larches offer solemn witness to the forest that was. Now, however, we are driving through a dense stand of small lodgepole pines, all nearly of a size, short, straight, and with only a short tuft of branches at the top. The lodgepole pine, like its first cousin, the jack pine of the Lake States, learned to make a blessing of adversity. It is poorly equipped to compete with other trees of an established forest. Its seeds, unlike those of the hemlock or grand fir, cannot sprout and establish a seedling on rotten wood and decaying vegetation. It can't endure shade, which accounts for its slender form with the branches only at the top where they can get sunlight. All of the lower limbs have been killed off by shade. Yet it is unable to grow as tall as the larch or fir and thus get sun. Therefore in old, mature, mixed forest, the lodgepole pine is a minor species, with only a relatively small number of individual trees hanging on in openings or on some rocky knoll. There it bides its time and does something few other trees can do; it accumulates a store of seeds. Its cones are small with their scales pressed tightly together and few open to scatter seeds where they would be likely to rot in the shade or be eaten by rodents.

Then comes a forest fire. Most of the trees are killed, including the lodgepoles. This is the time the lodgepole is prepared for and the destruction gives the species its opportunity. The other trees have already scattered their seeds to the duff of the forest floor where they have perished in the fire. But on every fire-killed lodgepole hang tight little cones clutching seeds produced for years back. Now, dried by the heat of the fire, they open and make their attack just when their competitors are helpless and bankrupt. Down flutter their seeds, each with a cleverly designed wing shaped to assure that even with little wind the seed will be scattered widely. Gone are the rotten wood and duff, and they find lodgment in mineral soil and ashes, their favorite seed bed. Frequently, as here, they really overdo it and grow so thickly that they crowd each other and need to be thinned as I thin my carrots. Nevertheless, where once the lodgepole was a despised slum dweller, it now rules supreme.

The other species have not surrendered; this is merely a truce. Over there is a tall larch that survived the flames. Already it is producing seed to rejuvenate its species. Along the watercourse, where dampness slowed the fire, some spruces and grand firs survive and they, too, are rallying to the attack with the advantage that they can grow under shade of the lodgepole. On rocky points, Douglas-firs with thick, fire-resistant bark are still living and producing seeds. For a hundred years or more these species will be where the lodgepole was before the fire, just hanging on to a foothold. But the lodgepole will reestablish precisely the conditions of shade and duff that handicapped it before. All the time the other species will be sending out raiding parties, establishing an outpost here, capturing a salient there, and infiltrating the main stand. Finally, as is the way with conquerors, the lodgepole will degenerate under its prosperity. It will become old and weak; insects, fungi, and wind will take their toll. Then the waiting invaders will rush in to reclaim their home, relegating the remaining lodgepoles to live again in subordination until fire gives them another opportunity.

There is much more to look at around here and through the flats between West Glacier and Lake McDonald, but perhaps we should continue on up to Avalanche Campground. Please note, though, that we are beginning to see some new species of trees along the lake and up to the campground. We are getting to an entirely new plant association. Here are western red cedar, white pine, and hemlock, also a few clumps of yew. Now, too, we begin to find the white spires of beargrass. That is another of those foolish names, for it isn't a grass and so far as I have been able to discover, bears care little for it. It's a member of the lily family.

At Avalanche we are in the heart of the new plant society. This is something you won't see in Yellowstone or in many other places this far east. Here, within just a few miles of the Atlantic slope, is an area that makes the botanist feel he is in the Cascade Range or at least in northern Idaho around Priest Lake. Right here and on the South Fork of the Flathead River are the furthest east projections of a Pacific slope vegetation type characterized by cedar, hemlock, yew, devil's club, mountain lover, ferns, and bracken. There is much the same type over at the DeVoto Memorial Cedar Grove along the Lochsa River in Idaho. There, nearer their true home, the trees are larger and more stately. In the evolutionary scale this is a more primitive, ancient, and less highly developed group of plants.

The white spires of beargrass in Glacier National Park. —Photo courtesy National Park Service

The yew and cedar are both relatively old and not highly developed representatives of the conifers. The ferns are the real members of the Society of Colonial Dames, however; they trace their family straight to the Carboniferous age. I often have wondered how this type came to be this far east. It seems out of place. When the glaciers withdrew, did it rush in from the west and usurp this little stretch of territory? Or was the type generally distributed throughout western Montana, only to be driven out of the drier areas? Probably some ecologists or paleobotanists have found the answer to this, but if so I have missed it. However it got here, it is probably headed for extinction.

Here, too, we find white birch. In general the broad-leaved, deciduous trees have not been very successful in invading the arctic, but the birches, quaking aspen, and some other poplars and stunted willows are exceptions, the birches outstandingly so. The birch survives by yielding to the weight of snow without breaking. Have you read Robert Frost's poem "Birches"? In it he tells of the boyish sport of swinging from the tops of birch trees: "So was I once a swinger of birches." Snow and ice were the original swingers of birches, but when released they usually spring back to their original position undamaged. The willows do the same thing but the aspens are less good at it. When they do break, they promptly send up shoots from the roots. All around the arctic across the northern part of this continent and into northern Siberia and Europe these species thrive even in permafrost where the ground thaws in summer to a depth of only a foot or two.

Some folks seem to require a miracle to assure their faith in the existence of a divinity. Perhaps here at Avalanche would be a good place to renew our faith and collect something to wonder at by looking at a miracle. Let's look at one of those tall larches; it must be 150 or 200 feet tall. Every summer day it must absorb through its roots several hundred pounds of water and lift it up through its trunk 150 feet where it is manufactured into starches and sugars (in combination with carbon dioxide from the air) or is transpired from the leaves. The miracle consists of the tree lifting such a weight to such a height. The transpiration of water from the needles creates a negative water pressure in the cells of the needle. The water is pulled, from one cell to another, all the way from the base of the tree.

If that is not a sufficient miracle for you, we can walk up to the gorge and look at another event at which I wonder. It does fairly well until a real miracle occurs. See how the trees and shrubs have roots penetrating the smallest cracks in the rocks, and more, they are actually forcing the rocks apart. Now consider the force necessary to split these rocks in two. This is roughly how they do it. They start with a little crack in the rock, one probably developed either by stresses and strains or by frost. Microscopic rootlets first penetrate the crack, then they start inserting molecules into their structure, growing, that is, a molecule at a time. In this process they develop pressures up to fifteen atmospheres or over two hundred pounds per square inch. Since there may be a total of tens and even hundreds of square inches of root surface in the whole mass of roots, the pressures build up to a total of tons. One doesn't have to come to Glacier National Park to see this phenomenon. All of us have seen concrete sidewalks and pavements lifted by tree roots. This is more of the process of reducing rocks to soil. After the lichens have etched little footholds, and the mosses and other small plants have built up a little organic material, up comes the heavy machinery, the woody plants, and the rock really is subjected to forces. Then water and ice will help to complete the quarrying job.

There are a lot of other things around Avalanche to see, but we shall take time only to look at that devil's club, which grows along the north side of the stream. Fortunately this plant does not grow much through our Montana Rockies. The other big concentration that I have seen in the state was in the area now flooded by Hungry Horse Reservoir. Its destruction is the only bright spot to me in the ruination of that area. In Idaho, Washington, British Columbia, and Alaska I have seen stream bottoms filled with a solid mass of it, sometimes as high as one's head. Notice how its stems reach out in a semirecumbent position and how every part of the stem and even the underside of the leaves are covered with slender, sharp spines. I never have seen a deer or even a bear attempting to travel through devil's club thickets. The spines will penetrate even heavy canvas trousers, and every one breaks off and lodges in the victim. In Latin it is variously classified as *Oplopanax horridus*, and "horrid" is fitting. However, I have heard it called a lot of other things equally appropriate and considerably more forceful.

When we leave Avalanche we begin to climb rapidly. Now we shall see the more striking effects of altitude, cold, and snow approach arctic conditions. As we start up the mountain and leave the valley, we are in what ecologists call the Hudsonian Zone, named, of course, for Hudson's Bay. We are leaving behind the larch, Douglas-fir, cedar, and hemlock. Lodgepole pine is becoming scarcer. All are becoming replaced by Engelmann spruce (which grows all the way from the lower to the upper timberlines), subalpine fir, and a five-needled white pine, the whitebark pine. These are true subalpine trees in this part of the country. Yew persists pretty well up to the switchback. But notice how the alder and snowbrush grow here, in dense masses of reclining stems all pointing downhill. In winter it will be completely covered by snow and pressed to the ground. An avalanche can pass over it with little or no harm to the plant. When summer comes, it will spring back and thrust up its tips, clothe them with leaves, and resume growth undamaged. Along the streams the alder becomes a small tree; here it adapts itself to its environment and assumes an entirely different habit of growth, so different that most botanists regard it as a separate species, though there is difference of opinion and disagreement in the classification of alders. Don't try to climb up one of these slopes covered with alder and snowbrush. You will find it nearly impenetrable.

Up near the tunnel, the Park Service has erected a sign calling attention to some fossil algae. The alga is one of the most primitive forms of plant life dating back to the earliest geological ages in which any life can be detected. Presumably it was from algae that most, if not all, other forms of plant life evolved. They occur as single, microscopic cells, as fine green filaments, and even as more elaborate plants in some of the seaweeds. The common green pond scum is made up of algae. It is an important element of the plankton, which is the basic food of all marine life. The importance of its role in the development and maintenance of the entire living world, plant and animal, can hardly be overestimated. We could have observed it alive in every little pond we have passed. Here are its fossil remains testifying to its antiquity. It survives alone or in combination with fungi where no other form of life can. I have seen banks of snow in summer, tinted a faint pink. When one disturbs it, it gives off a peculiar odor; some say it smells like watermelon juice. That pink color came from millions of single-celled algae in the film of water on the snow crystals. It is axiomatic that plant life preceded animals on this earth, since all animal life obtains its food directly or indirectly from plants. Since these algae represent about the most primitive and earliest form of plant, what we see here in rock is the beginning of living things so far as we know. This is the genuine prototype. If we adhere to the sometimes advanced theory that animal life was an offshoot of plants this may be the form of life from which you and I descended. So don't take the green pond scum lightly. Certainly this fossil merits at least the modest sign the National Park Service has erected.

Much of the timber along here has been killed by fire, but there are a few dense stands below the road. Notice that at this elevation the fallen twigs and branches rot very slowly. There is a thick duff under the timber. This, too, rots and turns into humus slowly for the same reason food doesn't spoil as quickly in the refrigerator as in the warm kitchen. As it slowly rots, it produces humic acid. Practically all of these alpine soils are acid. That just suits huckleberries, so we see them here too. That genus,

Vaccinium, is another that circles the northern hemisphere and is represented by several species, some of which differ from others by minor details only. In this locality there are at least two species, the ordinary one growing about two feet tall and the low one, which often forms an almost solid carpet. Other vacciniums are the cranberry, the lingonberry of Scandinavia and Alaska, the whortleberry of Europe, and of course the blueberries of the northeastern part of this continent. Some of the other plants that favor acid soil are the wintergreen, the laurels, the heath family, all of which are represented around here.

I regret I am not a zoologist so that we could study the animal life along the way. It too has been changing as we have climbed. Just for example, the jack rabbit of the plains was replaced by cottontails in the Flathead Valley which in turn have given way here to the arctic hare, or snowshoe rabbit. Presently we shall leave it behind and enter the habitat of the mantled ground squirrel, the whistling marmot, and the pika.

Animals face great difficulty at these elevations and resort to different expedients to solve their problems. Many rodents burrow tunnels under the snow and live like Eskimos in their igloos. The pika harvests and stores great quantities of hay for the winter. The marmot hibernates. Larger mammals migrate to lower elevations and by doing so get much the same climate effect as birds do by flying south. The elk likes to drop down into the open country and leave the forest. There it often raids haystacks and is a nuisance to ranchers. But then, the elk is naturally a plains animal and has adapted itself to the mountains only to escape persistent hunting. Probably there were few elk at this altitude before they were driven back here by white men. The pronghorn antelope refused to make this adjustment and because of its conservatism nearly became extinct. Fortunately protection came in time to save it.

Now it is time to drive on up to the top of the pass. On the way keep a watch for the pink monkeyflower along the watercourses. A member of the figwort family, it looks like a snapdragon. It is one of the most attractive of the alpine flowers. At lower elevations the yellow variety is more common but here the pink or red ones predominate. Along the Garden Wall there are few trees and what there are find it difficult to survive, what with shallow soil and avalanches. This just isn't a good site for trees, regardless of elevation. Look at the distant, more gentle slopes and note how the timber is becoming progressively smaller and more scrubby. If you could examine these stands closely you would find that the valley species have vanished entirely except for the Engelmann spruce which, as I have said, grows from top to bottom. But even it bears little resemblance here to the stately trees of the valley.

165

Alpine tundra above Logan Pass in Glacier National Park. —Photo courtesy National Park Service

Now we are at the top of the pass. There are still trees of a sort, but this is their upper limit. Perhaps you thought those were young trees growing up after a fire, but they are not young. The dead ones have simply died from age and the diseases that attack old, decadent trees. If you were to cut a cross section of the stem of one of these trees, you would find that you would need a strong lens to count the annual rings. Often they require fifty to one hundred years to make an inch of diameter growth. They have only a short two months each year in which to grow. Well up to the first of July they are buried in snow. Three species predominate, the subalpine fir, the persistent spruce, and the whitebark pine. In some places one finds larch, but it is a different species from that in the valley. This is subalpine larch, or Lyall's larch. It looks much like the eastern tamarack that farther north is found west as far as the Tanana Valley in Alaska.

Notice how the trunks of the trees often have burls and deformities. This probably is a pressure wood, a hypertrophy. When one side of a tree trunk or branch is subjected to pressure, an abnormal growth is stimulated, the result of the tree's effort to resist the pressure. These trees are constantly subjected to strains by the loads of snow and they rally their strength to resist, hence these strange deformities. I often have thought how like human beings are these trees. Like trees, people in poor economic circumstances and subjected to oppression tend to become warped and stunted both in body and character. Trees, however, cannot become antisocial or criminal; only their bodies are warped. Even so, timberline trees always remind me of impoverished humans who have merely responded to the pressures they have encountered.

What can one say of the flowers? They really defy description and a mere enumeration of species means little and does no justice to their beauty. I do want to point out some members of the heath family, for the heathers are northern plants. I have already mentioned the snowbrush, *Menziesia*. Kinnikinnick has a wide altitudinal range. The Labrador tea is evergreen and grows in moist places, fairly high up. I like to crush its dark green leaves and smell its spicy fragrance. My favorite member of the family is the mountain heath, *Phyllodoce*, with its evergreen leaves that somewhat resemble fir needles. It is a woody plant growing about five to ten inches tall and bearing conspicuous clusters of red blossoms. I admire its hardihood. I have seen it thrusting its tips through a late-melting snowbank and starting to bloom before more than one-half its height was uncovered.

Perhaps I should name a few of the more conspicuous flowers, the beargrass you already know and the yellow glacier lily. Other yellow flowers are most alpine buttercups and arnica. That taller one with an arrowhead-shaped leaf is groundsel, or *Senecio*. Along the edges of the snowbanks are white queen's cups and smaller spring beauties, the ones with the delicate pink veins in their petals. Perhaps, too, you may find a white dryad. Brilliant Indian paintbrush contributes most of the red, but look for the mossy carpet of the *Silene*, or moss campion, and the purple red of the heather. Fireweed climbs up here and contributes its share to the red shades. Tall forget-me-nots provide the pale sky blue and the lower-growing gentians the rich dark blue. The colors of these alpine flowers are clear and bright.

Let's climb up on the cliffs a little way where there is no plant life or soil, just barren rock. But is it barren? No, here is a scale, gray-green in color, and there is another with a yellow tinge. Over there in that little crevice in the rock some moss has lodged, and down in that larger pocket is a *Mitella* with little greenish yellow flowers, and a little sedge. This is the front line of the campaign the plant kingdom is waging to clothe the earth. These scales are lichens, a strange companionship of an alga and a fungus teamed up together. The single-celled alga has chlorophyll and manufactures food for the team. The fungus mycelium lends strength to the firm. This is not a single plant like a pine tree or a strawberry plant; it is a sort of firm or copartnership. Each member can be cultured separately. Neither could live here on the rock alone, so they have joined in a perfect symbiosis to invade an inhospitable region. The lichen is probably the slowest growing of plants. That little spot, hardly more than an inch across, may have been growing fifty years. There are many kinds of lichens, differing widely in form, appearance, and habitat. The "moss" that hung from the trees in the valley wasn't a moss but a lichen, *Usnea* by name.

The thing to remember is that every lichen is truly two plants, a fungus and an alga. Each reproduces separately from the other. Each is capable of life separate from the other. Together, they can do what neither could do alone. This lichen we are looking at has a unique characteristic. It secretes a complex and very powerful acid that can etch the hardest rock, even crystals of mica or quartz. The acid is so powerful that neither the fungus nor the alga can tolerate it within their cells, so the fungus synthesizes the acid outside its cells. Alcohols developed by the alga combine with the acid to form brilliantly colored salts. The crucial point is that with its acid, the fungus breaches the solid surface of the rock so that it will weather faster. At the same time it reduces the mineral matter to soluble form suitable for plant assimilation. It has produced a little niche, where its decaying organic components combine with mineral matter from the rock to form a small particle of soil.

Then, perhaps, a spore of moss lodges in this soil and, growing, forms more organic material and, consequently, more soil. Next comes some seed plant, perhaps a sedge or grass, and its little roots penetrate every minute aperture in the rock and aid the frost in breaking it up. We are now well on our way to planting a garden. In fact, those meadows brilliant with flowers are there partly because the lichens and mosses were capable of starting cultivation at these altitudes on these rocks.

Let me repeat that back in the Precambrian sea, the single-celled algae were one of the first forms of life. They learned to grow together in threads. The long process of evolution was started, from which came the most elaborate plant forms. Now here on these alpine rocks the algae, this time teamed up with that recessive renegade decadent of the plant kingdom, the fungus, is still patiently at work clothing the world in green. It is interesting to speculate what would happen if every living thing on the face of the earth were destroyed except one single colony of algae. Theoretically, that colony would have the potentiality to repopulate the earth with plant life and to develop again to complex forms. What would these forms be like? Certainly not like those we know now; but equally, certainly, they would be adapted to their environment, in perfect adjustment, for that is the law of nature. That life, which adjusts itself to the conditions it encounters, which can change and adapt itself to changing conditions, survives. The weaklings, and those that refuse to change as the world changes, perish.

Now it is time to leave these heights. We are not lichens and cannot get our nourishment from these rocks. But before we go, let's take one more look at the view before us. I am afraid that all too many looking at this same view see only a picture—a static, even dead, picture. True, they may find it beautiful and even inspiring, yet to them it is, after all, a picture different only in its depth and grandeur from a painting or photograph. What I see is a panorama of a dynamic and ever-changing world teeming with life even on the highest mountaintop. I see a divine plan in its execution.

INDEX

Page numbers in bold typeface refer to photographs or maps.

John B. Taylor was born on a farm in Nebraska in 1889, and his family moved to Missoula, Montana, in 1904. John Taylor was a trained conservationist, who was a part of the US Forest Service from 1907, two years after its inception, until his retirement in 1950. A man with a bachelor's degree in liberal arts, and a master's degree in botany and forestry, he was nevertheless a more-than-competent logger, cowboy, engineer, firefighter, and above all, humanitarian. He married Catherine Hauck in 1926 and together they raised three children, Elsie, Dora, and Ellen. He died in Missoula in 1975 at the age of 84.

In his acknowledgments, John Taylor thanked A. B. Guthrie and Harold G. Merriam for their encouragement and counsel. A. B. Guthrie was in the same graduating class (1923) at the University of Montana (then Montana State University) as Taylor's wife, and she may have introduced them. The book editor, John C. Frohlicher, graduated in 1926 and may also have known Guthrie, as both were involved with *The Frontier*, the university's literary magazine, under the guidance of Harold G. Merriam. A reknowned creative writing professor, Merriam was influential in turning Missoula into a hub of literary talent.

John C. Frohlicher was born in Kalispell, Montana, in 1901. He studied English at the University of Montana and was editor of *The Frontier*, the university's literary magazine. While at school in the 1920s, he worked summers for the Flathead National Forest, and much of his writing deals with his experiences in forestry and fighting fire. He went into journalism, working for the *Butte Standard* and then the *Missoulian*. He was engaged to John Taylor's sister-in-law, Elsie Hauck, but she died in 1928, before they were married. He moved to St. Paul, Minnesota, in the 1930s but stayed in touch with the Taylors and edited the memoir in the late 1960s. Frohlicher wrote that the hardest job he had as editor was battling Taylor's incredible modesty. Frohlicher died in 1988.

John N. Maclean, an award-winning author and journalist, has written about wildland fire for more than two decades. His most recent book, *The Esperanza Fire: Arson, Murder and the Agony of Engine 57*, is being developed into a feature-length film by Legendary Pictures and director Jim Mickle. His first book, *Fire on the Mountain*, a critically acclaimed account of the 1994 fire on Storm King Mountain that took fourteen lives, has been the basis of several television documentaries. Maclean was a reporter, editor, and writer for the *Chicago Tribune* for thirty years before he quit to write about wildland fire. He is a frequent speaker at fire conferences, academies, and other gatherings.